Iris Biometric Model for Secured Network Access

Iris Biometric Model for Secured Network Access

Franjieh El Khoury

CRC Press
Taylor & Francis Group
Boca Raton   London   New York

CRC Press is an imprint of the
Taylor & Francis Group, an **informa** business
AN AUERBACH BOOK

CRC Press
Taylor & Francis Group
6000 Broken Sound Parkway NW, Suite 300
Boca Raton, FL 33487-2742

Printed on acid-free paper
Version Date: 20130322

International Standard Book Number-13: 978-1-4665-0213-0 (Hardback)

Library of Congress Cataloging-in-Publication Data

Khoury, Franjieh El.
   Iris biometric model for secured network access / Franjieh El Khoury.
   pages cm
   Includes bibliographical references and index.
   ISBN 978-1-4665-0213-0 (hardback)
   1. Biometric identification. 2. Iris (Eye) 3. Computer networks--Security measures. I. Title.

TK7882.B56K46 2013
005.8--dc23
                                                                    2012051055

Visit the Taylor & Francis Web site at
http://www.taylorandfrancis.com

and the CRC Press Web site at
http://www.crcpress.com

Dedication

To my father,
I dedicate this work.

Contest all limits...

Contents

PART 2   CRITICAL ANALYSIS OF METHODS OF IRIS RECOGNITION

CHAPTER 5   EXISTING METHODS FOR LOCALIZATION OF EXTERNAL AND INTERNAL EDGES OF THE IRIS OF THE EYE

CHAPTER 6   EXISTING METHODS FOR ELIMINATING EFFECTS OF THE EYELIDS

PART 3   OUR PROPOSED MODEL: THE IRIS CRYPTO AGENT SYSTEM

CHAPTER 7   BIOMETRIC MODEL FOR AUTHENTICATION USING THE IRIS OF THE EYE

Acknowledgments

First, I would like sincerely to thank Mr. Rich O'Hanley, publisher of Taylor & Francis, for his patience and support.

The expression of my sincere gratitude goes to Dr. Fouad Ounsi for his advice, encouragement, and support.

I would like to thank my thesis director, Professor Marcel Egea and my codirector, Professor Nagi Wakim, for their guidance and advice.

I am grateful to the Institute of Automation, Chinese Academy of Sciences for allowing me to access its CASIA iris image database.

My gratitude also goes to the members of the research lab ERIC for their professional support, especially Professor Stephane Bonnevay and the members of the university USEK.

Thank you to Messrs. Bashir Bassatne, Eng. Joseph Geagea, and Ashraf Falou, who gave me all their support.

Thank you to Mr. Gilles Cattan, engineer and former director of R&D of SFIM industries, for his advice and moral and scientific support.

I also would like to acknowledge my friends who encouraged and supported me throughout my work. Thank you to Dr. Nada Meskaoui for networks, SMA references, and advice. I thank Dr. Mirna Abboud and Mr. Eng. Antoine Zgheib for their support and their encouragement.

I sincerely thank all my family for the sacrifices they made so that I could achieve my goal.

Introduction

The rapid evolution of technology in the computer world has made securing access to confidential data a very important issue in terms of research. This technology is likely to evolve in various architectures. Each architecture has its own functionality and its advantages and disadvantages. On the one hand, the transition from a centralized to a decentralized system distributed locally or remotely has facilitated the tasks in various sectors (e.g., educational, social, government, commercial, etc.). On the other hand, the integration of the Internet has encouraged communication at an international level and allowed operations to be conducted remotely (e.g., e-commerce, e-banking, etc.). In addition to the Internet, there is the mobile network that provides advanced services (e.g., multimedia messaging, Internet access, etc.).

This growth was followed by development of several methods of accessing data such as traditional (e.g., password, smart card, etc.) or biometrics (e.g., fingerprint, hand geometry, iris identification, etc.).

The complexity of computer networks, the variety of applications, and traditional methods of accessing security encountered difficulties regarding access controls to data and increased risk of attack of confidential information.

In order to provide a network's user a well-secured access level and guarantee the protection of confidential information, our study focuses on the development of the model IrisCryptoAgentSystem (ICAS), which can meet these objectives. This model is based on the biometric method of "the iris of the eye" (for authentication of individuals) and the method of asymmetric encryption (for data encryption).

We also propose to integrate a multiagent system (MAS) composed of agents of different types (e.g., biometrics, cryptography, etc.). The MAS system must be capable of meeting the multiexpertise present in our model.

Our work improves methods for the localization of external and internal edges of the iris of the eye. In addition, our work includes new concepts at the level of

- The classification of biometric signature of the iris to optimize the search time by introducing the concept of hierarchy indexed by trees and the pretopological aspects
- Removing the effects of upper and lower eyelids on the iris

Presentation of the Book

This book is composed of several parts divided into chapters to achieve a relevant model able to secure access to confidential information.

In the first part, we show in Chapter 2 how biometrics can be implemented to solve security problems in a complex system. Such a system based on these principles has several levels of security and several means of defense that are detailed in Chapter 1. On the one hand, this system is able to recognize various forms depending on the recognition process and encrypt the information through various cryptographic methods presented in Chapter 3. On the other hand, we reflect MASs in detail in Chapter 4 and the interest to use them in various fields through the integration of various features of agents.

In the second part, we present in Chapter 5 the various conventional methods based on five simulations for the localization of external and internal edges of the iris of the eye, and the effectiveness of each. In Chapter 6, we describe current methods for eliminating the effects of upper and lower eyelids based on three simulations.

In the third part, we propose our model, ICAS, based on biometric methods using the iris of the eye for the authentication of individuals, and methods of asymmetric cryptography to encrypt information. Our biometric model is detailed in Chapter 7. This model allows the authentication of individuals by the biometric characteristics of the iris "gabarit" (template). Moreover, this model is able to encrypt the scanned gabarit by the asymmetric cryptography method using the Rivest–Shamir–Adleman (RSA) algorithm. In Chapter 8 we present our global model. This model is developed in a MAS composed of agents of different natures.

The fourth part is devoted to simulation and implementation of our model in hospital services. The methods we propose to improve the algorithm of iris recognition are presented in Chapter 9. The operation of these proposed methods is illustrated with examples of simulations in Chapter 10. In Chapter 11, we present a detailed description of a proposed application that will be implemented in hospital services.

Finally, we end with a conclusion that summarizes all the accomplished work and recommendations that can be realized in the near future.

PART 1
STATE OF THE ART

In this part we will present the different risks of attacks and their means of defense for the security of access to the information in the network. We will develop the different biometric techniques that are considered more reliable than traditional methods, such as the usage of passwords or smart cards. We will illustrate the different methods used to enhance the algorithm of recognition of the iris of the eye. We will present the different techniques of cryptography for the security of exchanged information over the network. Our reflection aims also to present the technology of multiagent systems (MASs) and the interest in integrating them into different domains of application.

1

SECURITY OF COMPUTER NETWORKS

The objective of this chapter is to present the different models concerning the risks of different aspects of attacks: physical, data and data transmission, and network. These risks have led to development of effective means of defense. We will emphasize the protection of data and its transmission across the network to ward off unauthorized access.

1.1 General Overview on Different Risks for and Means of Computer Defense

Security is defined by Lanctot (1997) as "a set of conditions and ways of acting." This set allows the user of a computer to accomplish, without interference, the desired tasks. According to Guy Pujolle (2003), security is considered "an essential function of networks since we are not able to see directly the destination. Security has a very important role to protect equipment that requires complete isolation from the external world." According to the Illinois Federation of Teachers (IFT) (Jalix 2001), security means "an adequate protection of property and persons."

Computer security is also presented as the detection of unauthorized actions and their prevention by users of a computer system. Consequently, security is essential against unauthorized access to and the protection of information stored locally or transmitted across a network.

1.2 Level of Security and Risks

We distinguish three levels of security (Lanctot 1997):

1. The lowest level, or the physical level, is relative to the security of the equipment.
2. The security of data constitutes the second level.
3. The third level is relative to the transmission of data.

The risks, spread over these levels, can be of different natures. Among these risks we mention (Lanctot 1997) the following:

- *Interruption of the system,* such as the destruction of a physical part or the cutting of a communication line
- *Interception,* such as unauthorized access to information
- *Modification* of the content of a message transmitted over a network
- *Manufacturing,* such as inserting a wrong message on a network or adding a record in a file

1.2.1 Risks at the Physical Level

Among the risks at the physical level we mention the following:

- *Theft* and the unauthorized use of computer equipment
- *Occasional dangers* such as easy connection of a portable device to a system, reaching information while traveling, the mention of the name of the user with his password in the case of a computer and the theft of a laptop

1.2.2 Risks at the Level of Access to Data and Their Transmission across the Network

It is possible to classify the risks at the level of data and their transmission across the network into several categories as follows:

- *Passive attacks* against confidentiality represent free access to the content of a message sent by an e-mail, a telephone conversation, or a file sent to a destination and containing sensitive or confidential information.
- *Active attacks* against the integrity of the data affect the security of data transmitted. These data can be presented either as a message or as a selected field in a message. Such attacks can

be represented as a connection failure, duplication, insertion, modification, rearrangement, or repetition.

- *Unauthorized access* to the availability of data is to deceive the verification of permission and user rights in order to access private and confidential information.
- *Repudiation* against the good reception of messages is operated by a receiver consulting a transmitted message of the type of recommended letter—for example, in case of no claim of receipt of this message.

1.2.3 Risks at the Network Level

The risks at the network level are classified into three categories: *common attacks, overload on the network,* and *detection of online machines:*

- *Common attacks* are mainly targeted at servers. It is possible to have several common types of attacks: the imitation of a legal user network by social engineering using fake names or references, the use of default accounts for network access, or the failure to update passwords or security configuration; the automatic modification of the source address by the mystified *Internet protocol* (IP) to appear as one of the original packet; the overuse of services, such as control over a server; and fraudulent access to the administrator account causing such deep changes.
- *Overload on the network* is characterized by sending a large amount of data that will consume disk space or bandwidth, sending excessive traffic on a specific port number to override the memory part restricted to variables, and utilization of processor (CPU) by running a script to turn off or reboot the system (e.g., denial of service [DOS] attacks).
- *Detection of online machines* is a significant risk. It is done by scanning the active ports and determining the names of computers, servers, and the connected users.

1.3 Means of Defense

Several means of defense have been considered against all risks spread over the three levels of security (physical, data and transmission, network) cited in the previous paragraph. Before detailing these means, it is interesting to cite some examples of security breaches:

- An *attack* was made on the *site Yahoo* on February 14, 2000.
- The *virus* "*I love you*" was sent by e-mail and activated when the user opened the attached file in the message. This virus multiplied by accessing the address book of the user and sending copies of itself. In addition, it destroyed in the physical medium all images with the extension .JPG.
- The *virus* sent as a *message on a mobile phone* in 2006 to "press the buttons '#' and '90,' '#' and '09,' or '#' and '9' to get a prize." When the receiver executed the order, he wasted units from the calling card.

1.3.1 *Means of Defense at the Physical Level*

To address the problems of failure, multiple means of control have been developed to secure access to hardware and software. These controls can be implemented according to the following three elements (Stallings 1999): *identification* and *authentication, authorization,* and *verification.*

1.3.1.1 *Identification and Authentication of Users*   The identification and authentication of users aims to have a personal profile, a smart card, or biometric authentication for secure access:

- The personal profile is certified by knowing the password or personal information. The password is shown on three levels. The first level of the password is on start-up, and it is checked by the BIOS during the power-up of the machine. The second is at the level of the start of the operating system, which provides access to the system (e.g., Windows, Unix, etc.). Finally, the third ensures at the level of the application the authorization of access and the security of private information. Given the importance of the password, conditions and techniques of different natures will be applied to this password:
  - Have a long enough password; change the password frequently.
  - Password should not be visible on the tape.

- Avoid first names and family names, the names of towns and streets, known brands of teams and services, or keywords; do not write it somewhere; insert special characters (#, $, @, *, &, !, ^, etc.).
- Choose a word with a spelling error (pictchure).
- Choose two juxtaposed words (cupflower).
- Reverse the chosen word (rewolfpuc) or shift one position on the keyboard (alarms becomes in Querty mode *s;st,d*).
- The smart card is used to access systems.

The use of this card is not very reliable. For example, a person may give his card to his colleague, allowing him to mark fraudulently his attendance at the job.

- The biometric authentication is applicable for recognizing a person according to his biometric characteristics. These characteristics are divided into three main types: morphological (i.e., physiological or static), behavioral (i.e., dynamic), and biological. The first type is related to physiological characteristics such as fingerprints, hand geometry, retina of the eye, the iris of the eye, facial features, and the veins in the back of the hand. The second type analyzes, for example, a classical signature, voice, keyboard rhythm, lip movement, or gait. The third type analyzes molecular structures such as blood groups or the deoxyribonucleic acid (DNA).

1.3.1.2 Authorization Authorization aims to assign privileges and global or partial access permissions. These can be presented in different ways. For example, a computer is accessible to the public or only to an X; a directory or a file is accessible only by the administrator; a file is accessible for reading to a user X, and for modification or deletion to a user Y; information is accessible for reading to a user X or confidential to a user Y.

1.3.1.3 Verification This type of control works by applying verification procedures for access to different levels of the computer system. Such control shall include the periodic procedures of change of password and, for example, the right of the backup copy of information on various devices.

*1.3.2 Means of Defense at the Level of Data and Their
         Transmission across the Network*

This section presents various techniques and protocols that protect data or private information transmitted over a network. These techniques are *data encryption,* the *IPSec protocol,* and the *SSL protocol.*

1.3.2.1 Data Encryption According to William Stallings (1999), cryptography aims to be able to transmit information in a secure manner based on a sequence of operations:

- The sender encrypts its message according to a process established by the receiver or by itself.
- The message is sent via any transmission.
- The receiver decrypts the message using the reverse process.

The data encryption function is to ensure the security of access to information represented in the form of an encrypted message. This message should be encoded by a key that references an encryption algorithm (Tomko 1996), "pretty good privacy" (PGP) (Pujolle 2003), or Kerberos (Microsoft 2002; Pujolle 2003).

Encryption technology is used in several applications, such as sending an e-mail, the transmission of a telephone call, etc.

1.3.2.2 Internet Protocol Security at the Level of Network Layer of TCP/IP
In 1994, the Internet Architecture Board (IAB) prepared a report entitled "Security in the Internet Architecture" (Stallings 1999). The report asserted that the Internet requires a fairly high level of security. Especially with e-commerce, it is essential to have security mechanisms that ensure confidentiality of credit card numbers transmitted over the network. The protocol transport control protocol/Internet protocol (TCP/IP) is used to interconnect thousands of users who sometimes keep their communications secret.

To overcome these problems, the protocol Internet protocol security (IPsec) introduced security mechanisms at the level of the protocol IP, regardless of transport protocol (Pujolle 2003). These mechanisms ensure integrity, authentication, confidentiality, and key management. An unsecured IP traffic is carried on each local area network (LAN).

IPSec provides the means for protecting communications over LANs, private networks or public wide area networks (WANs), and the Internet (Pujolle 2003).

1.3.2.3 Protocol Secure Sockets Layer at the Level of the Application Layer of TCP/IP The protocol secure sockets layer (SSL) is software to secure communications over the hypertext transfer protocol (HTTP) or file transfer protocol (FTP) (Pujolle 2003). SSL is used to implement security at the level of the application layer of the TCP/IP protocol over the TCP protocol (Stallings 1999), as shown in Figure 1.1. Electronic signatures are used to authenticate both ends of the communication and data integrity.

This software was developed by Netscape for its browser. It is alternatively integrated in specific packets and in a transparent way for the applications. Netscape browsers, browsers from Microsoft (e.g., Microsoft Explorer browser), and web servers are equipped with SSL.

The role of SSL is to encrypt messages, using the technology of public key cryptography between a browser and the interrogated web server (Microsoft 2002; Pujolle 2003).

SSL has become more important in electronic commerce—to secure the transmission of credit card numbers—than the simple security of a web communication. This protocol was applied to the system secure electronic transaction (SET) (Stallings 1999). For example, it serves to ensure confidentiality in the payment and delivery of information

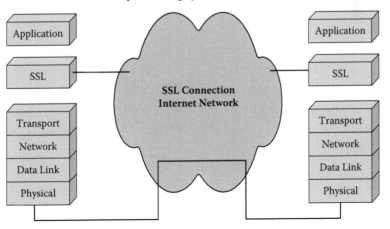

Figure 1.1   Architecture of SSL.

by encrypting them, to ensure the integrity of transmitted data using a digital signature, and to guarantee the authentication of the card-holder using the digital signature and the certificate.

Another protocol, secure hypertext protocol (S-HTTP) (Pujolle 2003), which is quite similar to SSL, has been developed to secure communications over HTTP, but it is less used.

1.3.3 Means of Defense at the Network Level

The protection at the network level is established by specific protocols and software to meet any kind of attack. This paragraph presents the protocol multiprotocol label switching (MPLS) and software of a type of firewall and antivirus used as means of defense at the network level.

1.3.3.1 Protocol Multiprotocol Label Switching The migration of the standard "frame relay" to the protocol MPLS gave more satisfaction, flexibility, and reliability in the transmission of information.

The protocol MPLS was proposed by the Internet standards organization (ITEF) (Pujolle 2003) for all architectures and higher level protocols such as IP, internetwork packet exchange (IPX), AppleTalk, etc. It has been introduced in several companies, such as Ipsilon, Computer Information System Company (Cisco), and aggregate route-based Internet protocol switching (ARIS) from IBM.

The standard frame relay and the protocol MPLS operate on the principle of packet routing IP following a path determined by a routing algorithm. However, the protocol MPLS is better than the standard frame relay since in the latter the route of the issued packet IP is traced in a static way across a number of predefined routers and each input reference corresponds to a single output. In this case, if any output port fails, there will be some delay in the transmission. This delay is due either to change routing of packets or to a connection failure. Consequently, it stops the transmission process until the node is recovered.

With the protocol MPLS, each input reference can have multiple outputs to take into account the multipoint addresses. Therefore, if an output port goes down automatically, the routing of the packet will be switched to a second without consequences such as connection failure or transmission delay.

The protocol MPLS contains specific transfer nodes called label-switched routers (LSRs) (Pujolle 2003). These LSRs behave as switches for the user data stream and as routers to trace the path with the signaling packet. In the case of the router, LSR participates in the implementation of the virtual circuit label switched path (LSP) through which the frames are forwarded. There is also in MPLS a mechanism of a stack of references to allow an LSP to transit non-MPLS or hierarchical domains through nodes.

1.3.3.2 Firewall The user needs to access to the Internet through the LAN or the Internet service provider (ISP), making the access of external world to resources of local networks easier. This creates problems of access control to resources and increases the risk of attack of confidential information. To overcome this problem, it will be necessary to find a practical approach such as the firewall to protect resources. This approach avoids applying to each machine very high standards of security features that require regular updating.

The firewall is a set of hardware and software (Microsoft 2002) to protect private network resources. It provides several protection services (Stallings 1999; Microsoft 2002; Pujolle 2003). The service network address translation (NAT) protects the addressing system of the internal network. The filters allow the flow of packets that belong to recognize the packets depending on the port numbers used in the applications. The service static address mapping hides the real internal addresses of accessible Internet resources.

In other words, a firewall is a specific router (Pujolle 2003) located at the entrance of a company whose goal is to prevent the entry to or exit from the company of unauthorized packets except those existing in a predetermined list (e.g., port 21 for ftp, port 80 for http, etc.). This is a drawback when a user of the organization connects to an external server to gather some information. In this case, the output by the firewall is accepted as it is authenticated. The answer is usually denied because the port on which this response occurs is blocked for security reasons. It will be necessary to authenticate this user by the server and to give permission to access the port, on which exists its response, by the firewall.

1.3.3.3 Antivirus   The antivirus is designed to detect the presence of a virus on a machine, identify its nature, and destroy it. However, the virus is not always up to detecting resistant viruses (Stallings 1999; Pujolle 2003). Some resistant viruses are not easily detected. The only means of defense for this kind of virus is regularly to update the antivirus used and not to open or execute a received message from an unknown sender.

1.4 Conclusion

We presented in this chapter various risks of attacks on three security levels: physical, data and their transmission, and network. We have also detailed several means of defenses adopted on these three levels. At the physical level, we discussed the identification and the authentication of users, authorization, and verification. At the level of data and their transmission, we presented the cryptographic methods and protocols IPsec and SSL. Finally, at the network level, we detailed the protocol MPLS and the firewall and antivirus software.

However, in a rapidly developed computer world, it is insufficient to adopt traditional authentication methods such as passwords or smart cards. It is this objective we develop in the next chapter—the means of defense using biometric techniques—in particular, the iris of the eye.

2

BIOMETRIC SYSTEMS

The development of biometric technologies aims to implement a method ensuring a high level of security. This chapter presents general biometric systems for data security and authentication of persons. A study of various biometric techniques is detailed. Then, a comparative study between these different techniques is established to specify which method is desirable. This comparative study shows that the method of biometric authentication based on the iris of the eye is most suitable for our research project. This justifies the presentation of the various algorithms for the recognition of the iris of the eye by stating, at each phase of the algorithm, the methods adopted. The aspects of discrete geometry for the definition of an edge in an image and the aspects of pretopology on the classification of objects are detailed at the end of this chapter. This is to introduce the concepts for the localization of the edge of the iris and the methods used for the classification of the iris in a database.

2.1 Introduction

Biometric techniques are known from the fourteenth century in which the Chinese put ink on the feet of children to identify them (Molineris 2006). In the nineteenth century, the French scientist Alphonse Bertillon was the first to call on the rich grooves of the inner ends of the fingers to identify offenders (Molineris 2006). The identification has entered the dictionary as a synonym for judicial recognition. After the attacks of September 11, 2001, the biometric technique has been implemented in authentication systems (or recognition) of users to prevent unauthorized access.

Despite the cost of tools, biometric authentication technology has become the most widely used technology. It is integrated in several sectors such as government sectors and the public and private sectors (Perronnin and Dugelay 2002).

This technology has demonstrated high reliability at the level of authentication. It ensures a high level of security to protect confidential information that is stored in a database or transmitted across a network. Moreover, it guarantees a very good control to secure private resources. This is compared to traditional methods (e.g., a password or smart card) that had notable weaknesses.

A password may be forgotten or guessed by someone else; as well, a smart card may be lost or given to another person, while biometric characteristics are innate and cannot be lost or stolen.

2.2 General Definition of a Biometric System

The use of a biometric system is done in three steps (Hashem 2000):

1. Acquisition of the image (e.g., take biometric measurements)
2. Extraction of relevant parameters
3. Identification (e.g., hiring a new person) or verification (e.g., authenticating access by a comparison with data stored in a database)

The effectiveness of this system is related to its adaptation to permanent changes in the user state (e.g., having a beard, wearing glasses, etc.) and variations in the ambient environment (e.g., effect of illumination, noise, etc.) (Gillerm 2007).

The performance of a biometric system can be measured primarily using three criteria: accuracy, efficiency (i.e., speed of execution), and data volume. Biometric techniques offer levels of security and facilities of different employment. Several factors are used to study a biometric system (Gillerm 2007). Among these factors we can mention:

- The cost of this sensor technology, which affects the total price of the system
- The response time of imposing the field of use, which reflects the effectiveness of the system (e.g., a rapid response system is considered more efficient if there is a high flow of users, as in airports)

- The reliability and certainty of the answer, which improve system security
- The environment of use (e.g., telephone for voice recognition, camera for authentication of a person using the iris, etc.)

In addition to these factors, three terms are considered as essential to clarify the performance of a biometric system (Gillerm 2007):

1. The false rejection rate (FRR) is the probability that the biometric system fails in the authentication of a registered person (i.e., rejects a real user or a valid ID).
2. The false acceptance rate (FAR) represents the probability of accepted verifications by error (i.e., provides access to an impostor). This depends on the quality of systems as well as the level of desired security.
3. The equal error rate (EER) gives the value when the FAR and FRR are equal.

Therefore, a biometric identification system is ideal where FRR = FAR = 0 (Mahmoudi 2000).

To ensure a high level of security, it is necessary to have a value of FAR that is very low. A compromise should be made in choosing the threshold to achieve a desired and preset value of FRR or FAR. This threshold depends on each application.

2.3 Different Biometric Technologies

Biometric techniques are increasingly presented as a way to fight against fraud and theft (Mordini 2005). Recently, scientific methods for identification of people using biometric data were subjected to rapid and gradual technological evolution (Cabal 2003). These methods are classified into three categories (Perronnin and Dugelay 2002):

1. *Morphological,* such as fingerprints, hand geometry, retina of the eye, iris of the eye, facial features, and veins in the back of the hand
2. *Behavioral,* such as biometric signature, voice issues, rhythm on the keyboard, lip movement, and gait movement
3. *Biological,* such as deoxyribonucleic acid (DNA) blood groups

2.3.1 Morphological Biometrics

Morphological biometrics is related to physical characteristics. It is static and not easily modified.

2.3.1.1 Fingerprints   The fingerprints of each person are identified by the following:

- The patterns of the ridges (Figure 2.1), which are the bifurcation, the center, and the ridge end (Escobar 2006)
- The features (Jain, Ross, and Prabhakar 2004), which are represented as a snail, an arch, or a loop (Figure 2.2)

2.3.1.1.1 Fields of Application   This technique is used most in the market. MasterCard estimates that about 80% of transactions are performed by systems with fingerprint recognition (Perronnin and Dugelay 2002). The Federal Bureau of Investigation (FBI) and the Department of Homeland Security (DHS) have adopted the biometric fingerprint national fingerprint image software 2 (NFIS2) (Cho, Chande, and Li 2005). Moreover, this technique is used in accessing

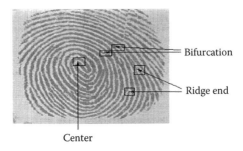

Figure 2.1   Patterns of ridges.

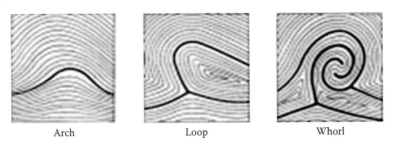

Figure 2.2   Features of the fingerprint.

controlled laboratories, premises that contain servers, and areas that require a high level of security (FaceKey Corporation 2007). As well, it is used for access to banks (e.g., withdrawing money at an ATM or payment by card) and for civilian applications (e.g., mobile or tele-working e-commerce) (Perronnin and Dugelay 2002). It constitutes a factor that makes a security service suitable for portable devices, applications of information technology, government, aviation security, and programs to reduce fraud (Michael 2002). In addition to the usual access methods, this technique is integrated into the encryption process so that the key cannot be seen without biometric authentication success (Uludag et al. 2004).

2.3.1.1.2 Advantages and Disadvantages The biometric fingerprint is well known. It provides a level of security at a good price (Michael 2002). It frees people from their password or PIN use. It is fast and easy to use. It is possible to change the authentication of a finger to another.

On the other hand, this technique has several disadvantages. It has a high error rate that is approximately $1/10^2$ (Rosistem 2001). It is difficult for some users, given the small size of the reader (i.e., some users have fingers sized larger than the size of the reader). In addition to the reader size, the need for correct position of the finger on the reader needs the cooperation of the user. Some systems accept a deformed finger (i.e., small cut or wound). As well, a small wound can cause a problem.

2.3.1.2 Hand Geometry This technique is characterized, as shown in Figure 2.3, by the hand size, the length and the width of the fingers, and the joints and their relative locations (Jain et al. 2004).

2.3.1.2.1 Fields of Application Hand geometry is a physical biometric technology (Escobar 2006). Over 90% of US nuclear centers use this technique, as well as the US military (Chasse 2002). Moreover, in the United States, hand geometry has been adopted in schools, hospitals, cafeterias, daycare centers, prisons, and banks. This technology is adopted to control access to sensitive areas where large numbers of people travel (Gillerm 2007). It is applied, for example,

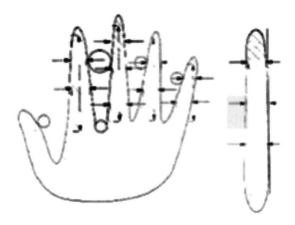

Figure 2.3   Measures of the geometry of the hand views above and to the side of the hand.

at the Olympics, at borders, in airports, and in major theme parks (e.g., Disney).

2.3.1.2.2 Advantages and Disadvantages   The authentication result of biometric technology based on hand geometry does not depend on humidity and cleanliness of the hand. The sensors of the hand geometry provide a reasonable level of accuracy (Gillerm 2007).

On the other hand, this technique has a wide reader, making it difficult to use by certain users (e.g., children, people with arthritis or missing fingers). The error rate is high for twins or members of the same family (Rosistem 2001), which is $1/7 \times 10^2$. Furthermore, the shape of the hand changes with age, which also influences the error rate.

2.3.1.3 Retina of the Eye   This technique is based on the biometric characteristics of the inner layer of the eye known as the retina (Escobar 2006). These characteristics are the features of blood vessels and veins (Figure 2.4).

2.3.1.3.1 Fields of Application   This technology is used in cases where safety is paramount. It is integrated in the military field, in the space sector—for example, in the National Aeronautics and Space Administration (NASA)—and by intelligence agencies such as the Central Intelligence Agency (CIA) (Chasse 2002).

Veins

Capillary

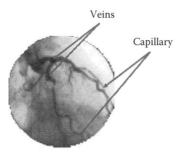

Figure 2.4 Characteristics of the retina of the eye. (Source: Bron, A. J. et al. 1997. *Wolf's anatomy of the eye and orbit.* London: Chapman & Hall Medical.)

2.3.1.3.2 Advantages and Disadvantages This technique is very reliable, given the distribution of blood vessels that is unique to each person, including twins. In addition, the retina cannot usually sustain injuries or burns.

As a disadvantage, this technique poses problems when scanning the iris for authentication because a single movement causes a rejection by the system. Moreover, this technique requires the use of a very sophisticated camera (Marie-Claude 2003). The appearance of the vessels may be somewhat modified by age or illness, but the relative position of the vessel remains unchanged throughout the life of the individual. On the other hand, the retina is influenced by the laser beams.

2.3.1.4 Iris of the Eye This technique measures only the characteristics of the annular colored region of the eye called the iris (Deluzarche 2006) (Figure 2.5). These characteristics are distributed in two

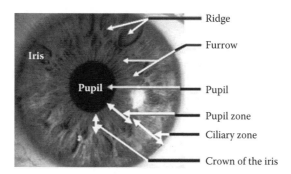

Figure 2.5 Characteristics of the iris of the eye. (Source: Bron, A. J. et al. 1997. *Wolf's anatomy of the eye and orbit.* London: Chapman & Hall Medical.)

different areas: the pupillary area and ciliary area (Bron, Tripathi, and Tripathi 1997). The pupillary area contains the crown of the iris. The ciliary area contains the furrows, ridges (or dots), and freckles.

2.3.1.4.1 Fields of Application   This technique is used, as a fingerprint is, in physical and logical security. Several applications of verification of iris recognition have been developed. The code of iridian, proposed by Daugman, has been used in many companies to produce their own products such as Panasonic's Authenticam (Michael 2002). This product is for private use of ID for iridian technology for iris recognition with security software of input/output. This technology allows multiple users access to computers, files, directories, applications, passwords of banks, and airports. Moreover, it helps to authenticate users who will access applications manipulating standard information (e.g., video conferences and online collaboration).

2.3.1.4.2 Advantages and Disadvantages   The biometric method of iris recognition is more efficient and accurate than other current methods used for securing access to data, since the error rate is minimal—in the range of $1/1.2 \times 10^6$ (Rosistem 2001). Moreover, iris patterns are developed during the first two years of life and are stable (Perronnin and Dugelay 2002). According to Bron et al. (1997), the irises are unique. The two irises of the same person are different, even for identical twins. The iris is not influenced by contact lenses or glasses. As well, it is not affected by cataracts or age.

On the other hand, the cost of biometric equipment of the iris scan is very high compared to that of other technologies. Furthermore, this technique is not practical in the case of use in a crowded area. As well, some people fear that their health status is also detected.

2.3.1.5 Face Recognition   This technique measures certain facial features (Escobar 2006). These features are the distance between the eyes, the distance between the eyes and edges of the nose, the angle of the cheek, the tilts of the nose, the thickness of lips, and the temperature of the face (Figure 2.6).

2.3.1.5.1 Fields of Application   This technique has been adopted in all international US embassies and government agencies (Escobar

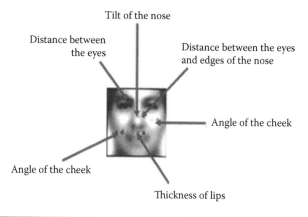

Tilt of the nose

Distance between the eyes

Distance between the eyes and edges of the nose

Angle of the cheek

Angle of the cheek

Thickness of lips

Figure 2.6 Characteristics of the face.

2006). It is also integrated into ATM machines and casinos to identify users, as well as in cities (e.g., Gold in Florida) to monitor citizens on public streets. Furthermore, it is used for one-to-many searches, verifications, inspections, and surveillance, as well as to ensure uniqueness in obtaining an image of the database (Michael 2002).

2.3.1.5.2 Advantages and Disadvantages FaceKey technology showed that the identification by face recognition includes two tasks: identify and recognize the face (FaceKey 2007). This technology works well with any race. It approaches the human method. It has a rapid detection.

On the other hand, this technology is affected by the appearance (e.g., beard, mustache, etc.) and environment (e.g., lighting or camera position) (Perronnin and Dugelay 2002). It requires several positions and expressions of the face to get good accuracy. In addition, this technology does not exclude a possibility of invasion by imitation.

2.3.1.6 Veins of the Palm of the Hand This technique takes into account the characteristics of the veins of the palm of the hand (Noisette 2005).

2.3.1.6.1 Fields of Application This technique still resides in the field of research. It was applied by a sector of the Japanese electronics group, Fujitsu Europe, as a new authentication technique (Noisette 2005).

2.3.1.6.2 Advantages and Disadvantages   The network of the palmar venous arch is unique to each individual, even in the case of identical twins (Noisette 2005). It can be recognized only with living individuals, when red blood cells circulate in the veins.

Unlike student systems using fingerprints, to identify it is necessary to have direct contact with the skin by placing the palm of the hand above the reader (Noisette 2005).

2.3.2 Behavioral Biometrics

Behavioral biometrics is dynamic. It varies with age or the behavior of the person.

2.3.2.1 Biometric Signature   The scanning of the biometric signature measures the speed, the acceleration, the angle of the pen, the pen pressure, the movement, the stroke order, the geometry, the signature image, and the points and time intervals where the pen is lifted (Jain et al. 2004; Escobar 2006).

2.3.2.1.1 Fields of Application   This technique is used in several countries as a legal or an administrative element (Gillerm 2007).

2.3.2.1.2 Advantages and Disadvantages   The process of generation of the signature is a reflex to exercise in real time and it is difficult to imitate, since the signature is a unique gesture to each individual (Jain et al. 2004). The advantage of this technique is that it is legally accepted. It is not physically stored.

On the other hand, this technique has a disadvantage in the variation of the signature for the same person depending on his or her state (e.g., tiredness, age, etc.). This creates a difficulty in achieving a highly accurate identification (Gillerm 2007).

2.3.2.2 Voice Aspects   The scanning of the voice measures the sound waves of human speech (Escobar 2006; Gillerm 2007) (Figure 2.7). The speaker recognition technology verifies the identity of the speaker by the spoken language and acoustic models (Jain et al. 2004). The speaker models reflect an anatomy. This anatomy is represented by the size and shape of the mouth, the size and shape of the groove, the

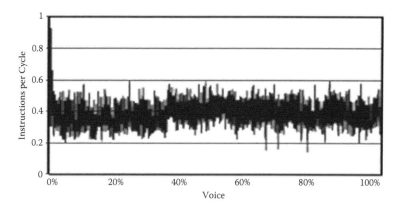

Figure 2.7   Instructions per cycle of the voice.

behavior (e.g., the tone of voice and the style of speaking), and, finally, by a signal requesting a heavy treatment (i.e., spectral analysis and frequency) (Chasse 2002).

2.3.2.2.1 Fields of Application   Voice recognition is used in popular applications to secure data over a telephone line (Phillips et al. 2000) or for the recognition of individuals (e.g., Sphinx-3 system developed by Carnegie Mellon University [CMU]) (Cho et al. 2005). This technology is also used in the verification of cases of house arrest (Escobar 2006). Where appropriate, at any time a computer calls a person at his home and he or she must answer the phone by saying a passphrase for authentication. As well, this technology has been adopted by police forces, intelligence agencies, immigration services, and hospitals (Gillerm 2007).

2.3.2.2.2 Advantages and Disadvantages   In this technique, the voice is the only information available in case of telephone transaction. The user can be located remotely. The voice is unique to each person.

On the other hand, this technique has a limitation in applications and a medium level of security. This is due to the high level of change of the voice of the individual. These changes are influenced by instability of health, emotional state, age, and environmental noises (Gillerm 2007). Furthermore, the possibility of imitating the voice of another person makes this technology useless.

2.3.2.3 Rhythm on the Keyboard   This technique consists of analyzing the typing speed, sequence of letters, typing time, and pauses for each person (Gillerm 2007).

2.3.2.3.1 Fields of Application   This technology is rarely applied.

2.3.2.3.2 Advantages and Disadvantages   This technique is very easy to install. It is unique to each person.

On the other hand, this technique is not accurate since the parameters are a function of the person's emotional state (e.g., tiredness) and physical condition (e.g., discomfort, illness, etc.) (Gillerm 2007).

2.3.2.4 Movement of the Lips   The model proposed for the modeling of the lips is composed of at least five independent curves (WIPO 2008). Some of these curves describe a part of the outer labial edge and at least two inner edges. The characteristic points of the mouth are analyzed using jointly discriminant information. This latter combines the luminance and chrominance, as well as the convergence of a type of active edge. It eliminates the parameter settings of the edge and its high dependence on the initial position.

2.3.2.4.1 Fields of Application   This technique is applied in the fields of video processing in real time. It has been used to improve speech recognition (i.e., reading lip movements) and to control an application for viewing panoramic images (i.e., combination of eye movements and voice commands) (Yang et al. 1998). As well, it is integrated in multimodality for security (e.g., automatic detection of playback) (Bonastre 2005).

2.3.2.4.2   Advantages and Disadvantage   In comparison to the previous behavioral biometric techniques, the footprint of the lips varies depending on the position of the mouth. The difficulty of the search for points, on the outer and the inner edges, is the fact that there may be various areas between the lips. These areas have characteristics (e.g., color, texture, or luminance) similar to or completely different from those of the lips when the mouth is open (WIPO 2008). Indeed, in conversation, the area between the lips can take different configurations (i.e., teeth, oral cavity, gums, and tongue).

2.3.2.5 Gait Movement   This technique measures, for each person, several different movements by jointed articulation. The process of gait recognition is based on four parts: the limit of the frame around the people that move, the extraction of the silhouette, the detection of the period of the step, and the estimation of similarity (Cho et al. 2005). This technique analyzes the image sequences based on speed, acceleration, and body movement (Deluzarche 2006).

2.3.2.5.1 Fields of Application   Applications are integrated to the recognition of gender of the person in question (Cho, Park, and Kwon 2003), as well as the gender classification (Lee and Grimson 2002; Yoo, Hwang, and Nixon 2005).

2.3.2.5.2 Advantages and Disadvantages   The level of performance of this technique remains in question, since recognition of a person is based on his way to walk and move (Cho et al. 2005). Therefore, this technique is affected by the physical state (e.g., illness, etc.) of the person.

2.3.3 Biological Biometrics

Biological biometric technology analyzes the molecular structures. It is not yet effective in terms of response time.

2.3.3.1 DNA   This technique is based on the genes encoded in DNA (Cite-Sciences 2005). These genes are the material carriers of heredity.

According to the laboratory of DNA testing specializing in genetic analysis in Spain (Neo Diagnostico 2008), the results are rendered in 3 to 5 working days, but in urgent cases these results are made available in 24 hours.

2.3.3.1.1 Fields of Application   The concept of DNA was introduced in 1985 by the English biologist Alec Jeffreys for the recognition of individuals (Gillerm 2007). It has been integrated to solve criminal cases (e.g., in 2003–2004, 43% of crimes in the world were solved using DNA evidence).

2.3.3.1.2 Advantages and Disadvantages The genetic information of an individual is unique since no member of a species has the same combination of genes encoded in DNA (Gillerm 2007). The analysis of DNA is a highly accurate method of identification that analyzes the evolution of molecular biology of a person.

On the other hand, this technique cannot be applied in real time, given the duration of treatment to identify an individual.

2.3.3.2 Blood Groups This technique characterizes the blood of each individual. These characteristics are the blood groups (i.e., A, B, AB, or O), the rhesus group (i.e., plus or minus), and others.

2.3.3.2.1 Fields of Application This technique has been adopted in the medical field to determine the behavior of an individual (Bourdel 1962; Montain 1999).

2.3.3.2.2 Advantages and Disadvantages This technique is unique to each individual, since no one is identical to another. On the other hand, it is necessary to take blood from the individual to be identified.

In addition to the biometric techniques detailed before, we can mention others in terms of research. These techniques are the geometry of the ear, drawing of the lips, body odor, heartbeat, teeth, analysis of the pores of the skin, saliva, blood flow, and many others.

2.4 Comparison of the Different Biometric Techniques

Table 2.1 and Table 2.2 represent a comparative study between different morphological and behavioral biometric techniques, as well as the strengths and the weaknesses of each (Bron et al. 1997; Perronnin and Dugelay 2002; Gillerm 2007).

Table 2.1 Comparison of Morphological Biometric Techniques

TECHNIQUES	STRENGTHS	WEAKNESSES
Fingerprint	• Possibility to change authentication from one finger to another • Quick and easy to use • Good accuracy	• Counterfeiting possible • Difficult to be accepted by certain users • Small wound causes a problem

(continued)

Table 2.1   Comparison of Morphological Biometric Techniques (continued)

TECHNIQUES	STRENGTHS	WEAKNESSES
Hand geometry	• Acceptable by users • Counterfeiting is in question	• Has a wide reader • Difficult to use by certain users (e.g., children, etc.) • Fairly accurate
Retina of the eye	• No infringement • Different characteristics for identical twins • Features are protected against changes in the external environment • More accurate	• Influence of laser radiation on the health of the user • Difficult to use
Iris of the eye	• Real twins have different irises • Not affected by age or by cataracts • Not affected by the glasses and lenses • Not changed by surgery • Rapid detection • Very accurate	• Intrusive • Some people worry that their health status is detected
Facial features	• Closed to the human method • No intruder • Rapid detection	• Requires multiple positions and facial expressions to get good accuracy • Affected by the appearance and environment • Possibility of invasion by imitation

Table 2.2   Comparison of Behavioral Biometric Techniques

TECHNIQUES	STRENGTHS	WEAKNESSES
Signature	• Difficult to imitate • Widely accepted • No intruder • No record of the signature	• Difficult to use • We must train the system • Inconsistency of the signature • Not very accurate • Takes into account the dynamics of the gesture and the appearance of the signature
Voice	• Easy to use • Acceptable by users • No intruder	• Counterfeiting easy • Attack prerecorded • Affected by ambient noise • Less accurate
Rhythm on the keyboard	• Continuous monitoring as the keyboard is used	• Not applicable for touch screens • Sensitive to the state of health of the person • Not yet well developed • Less precise

2.5 Algorithms for Iris Recognition

The systems of iris recognition are used to extract the biometric characteristics of the iris (i.e., biometric template of the iris or iris biometric signature) "gabarit" (template) for secure access to confidential data or private locations. These systems, like other biometric systems, consist first in enrolling the gabarit in a database of identification. Then they are used to create the gabarit for each person wishing to access the system. The required gabarit will be compared with other gabarits stored in the database to validate access.

The process of recognition by the biometric method of the iris is composed of six phases: acquisition of the image of the iris, manipulation of the image, localization of the external and internal edges of the iris, normalization to isolate the region of the iris from the image, extraction of the biometric characteristics of the iris, and identification or verification.

The first full biometric authentication system based on iris recognition was proposed by John Daugman (1993). This last one has developed the code of the iris "IrisCode" at Cambridge University (Potel 2002). This code is known as "iridian technology." The algorithm of Daugman provides 3.4 bits of data per square millimeter from an iris with a diameter of 11 mm. This implies that each iris has 256 unique points for traditional biometric technologies.

Several researchers have worked on the authentication process using the iris. They based their algorithms on the implementation of John Daugman. They proposed innovative techniques to improve the performance of specific phases in the recognition process (e.g., effect of the eyelids, problems of edge detection, etc.) as mentioned in the following sections.

2.5.1 Acquisition and Manipulation of the Image

The image acquisition consists of having an image in JPEG format with good quality. The image quality is influenced by the type of camera, the distance between the camera and the eye, the illumination intensity, and the noise.

Vatsa, Singh, and Gupta (2004) proved that one must have a distance between the CCD camera and the eye of 9 cm to have an image with good biometric characteristics of the iris. Tian, Li, and Sun (2006) used two different CCD cameras, with the same resolution, respectively, with a distance 4–30 cm and 70–90 cm between the eye and the camera. This gives flexibility in image acquisition.

The manipulation of the image aims to reduce the effect of illumination and noise, as well as to accelerate the identification process (i.e., reduce the size of the image). Ya-Huang et al. used the median filter (Huang, Weiluo, and Chen 2002). This filter is used to eliminate image noise with little appearance of the edges. However, the Gaussian filter has been adopted by Tian et al. (2006), Meng and Xu (2006), and Daouk et al. (2002). This filter helps to eliminate the noise and the effect of illumination and, consequently, to improve the identification rate. The most proposed algorithms reduce the image size to a quarter of the initial size and then convert this reduced image from RGB (red, green, blue) to grayscale level (Daouk et al. 2002; Schonberg and Kirovski 2004).

2.5.2 Localization of the External and Internal Edges of the Iris

The localization phase consists of locating the external edge of the iris and detecting the position of the pupil to locate the internal edge of the iris.

Indeed, Daugman used the integrodifferential operator to locate both external and internal edges of the iris (Huang et al. 2002). This operator reduces the speed of the overall research. As with Daugman, Ya-Huang et al. applied an integrodifferential operator, but they reduced the image size to decrease the complexity at the level of research time (Huang et al. 2002). In addition, they used the Canny operator to detect edges in the grayscale image. This operator is considered an optimum detector for edges corrupted by white noise (Christodoulou and North 2004). This operator is preceded by the Gaussian filter (Christodoulou and North 2004) to eliminate noise and thus make the characteristics of the iris more readable.

Wildes (1997) and Daouk et al. (2002) have adopted different methods based on the circular Hough transform. This transform is widely used (Narote, Narote, and Waghmare 2006). It is considered the best method for detecting edges of objects. This method consists in finding, in the reduced image to one quarter, the circle that corresponds to the largest circle containing the summation of maximum intensity for the external edge of the iris. Thus, this method uses a threshold (i.e., the highest threshold is used to the iris with clear color, for example, blue or green) for the Canny operator. This threshold depends on the intensity of pixels in the iris to locate the internal edge.

However, Tian et al. (2006) have adopted the search method of a square of 60×60 pixels in the image, to detect the position of the pupil. The research method consists of scanning adjacent squares to identify the one that has the lowest average of gray intensity (i.e., 0 = black and 255 = white). After the localization of the position of the pupil, they used the Canny operator to obtain a binary image. On this binary image, they applied the circular Hough transform to locate the internal edge of the iris. For the location of the external edge, they used the operator integrodifferential.

Meng and Xu (2006) adopted the Canny operator for edge detection. They searched the gray points, identifying the curve of the histogram of the iris, to solve the problems of the localization of the internal edge of the iris. They used the circular Hough transform for the localization of the external edge. However, Miyazawa et al. (2005) have located both external and internal edges by the elliptical Hough transform, which is considered a generalized case of the circular Hough transform. In addition, in their tests done on the database CASIA (2006), they adopted the angle θ equal to zero, which results in no degradation in performance. This method cannot be generalized since it is tested on specific images of the iris.

2.5.3 Normalization

The iris is captured in two different sizes for different people, as well as for the same person. This is due to several factors (e.g., variation of illumination, change in distance between the camera and the eye, eye position). If any, normalization is necessary because the size difference between the two compared irises affects the result of verification. Moreover, the normalization consists of reducing the distortion of the iris due to movement of the pupil (Meng and Xu 2006). It serves to break the nonconcentricity between the iris and pupil. It also aims to simplify the processes that follow.

The normalization consists of transforming the region of the iris, which is represented in the form of rings, to a rectangular shape with fixed size (Meng and Xu 2006; Xu, Zhang, and Ma 2006). This case requires a transformation of Cartesian coordinates to polar coordinates. Recall that each point making up the region of the iris is characterized by its coordinates (x, y) in Cartesian theory and its intensity of gray level $0 \leq \text{gray} (x, y) \leq 255$. Passing to polar coordinates, we obtain the following system:

$$0 \leq \theta \leq 2\pi$$

$$\rho_{\text{pupil}} \leq \rho \leq \rho_{\text{iris}}$$

$$\text{Gray}(\rho, \theta) = \text{gray}(x, y)$$

where ρ_{pupil} is the radius of the pupil, and ρ_{iris} is the radius of the iris.

We obtain a frame "raster" whose lines are marked by ρ and columns by θ. We illustrated this method by the graphic shown in Figure 2.8; the graphic in Figure 2.8(a) relates to the representation in Cartesian coordinates and the graphic in Figure 2.8(b) illustrates the representation in polar coordinates.

2.5.4 Extraction of Biometric Characteristics of the Iris

The extraction of the biometric characteristics of the iris gabarit consists of extracting from the region of the iris its parameters representing the texture information of the iris. As with Daugman, Meng and Xu (2006) and Xu et al. (2006) adopted the Gabor filter. This filter contains several attributes suitable for extracting information from iris texture.

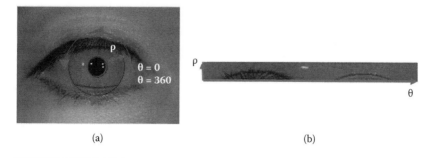

(a) (b)

Figure 2.8 Representation of the iris in (a) Cartesian and (b) polar coordinates.

Tian et al. (2006) used the two-dimensional (2D) zero-crossing detector. This filter helps in extracting the characteristics of the iris by calculating the convolution G (i.e., G = [–1 2 –1; –1 2 –1; –1 2 –1]). The result of the convolution is easily influenced by the illumination. But as the sign of the convolution result is stable, they are coded in binary to compose the binary model of the iris or "binary gabarit" of the characteristics of the iris. Tian et al. took the threshold value of 0 for the binary encoding of the gabarit. In other words, the binary method consists of adjusting the density of a pixel to one if its value is greater than or equal to zero, and this density is adjusted to zero if its value is less than zero.

However, Lim et al. (2001) adopted 2D discrete Haar transform. Daouk et al. (2002) adopted the same method as Lim et al. but at five levels of iterations. This transform has shown 96% efficiency. It is applied at five levels to reduce redundancy and minimize the size of the model extracted from the iris (Daouk et al. 2002).

2.5.5 Verification "Matching"

This phase aims to check whether the person is authorized to access the system. This verification is done by testing the conformity between the required model and the one stored in the database.

Miyazawa et al. (2005) adopted the correlation band-limited phase-only correlation (BLPOC) in the discrete Fourier transform in two dimensions to calculate the similarity between two templates. This method encounters a difficulty in selecting the threshold level for defining an acceptable correlation (i.e., based on the correlation between zero and one, the similarity of the gabarit will be graded on this scale).

As with Daugman (2004) and Wildes (1997), Huang et al. (2002), Robert, Bradford, and Delores (2005), Meng and Xu (2006), Xu et al. (2006), and Daouk et al. (2002) used the Hamming distance to check the dissimilarity between two binary templates with a threshold equal to 0.32. We will see that this method leads to some difficulties such as the justification of the selected threshold.

However, Tian et al. (2006) adopted the concept of vector verification. This method consists in considering the required model of the iris and the model stored in the database as two vectors. Their similarity is calculated by the cosine of the angle between their vectors, on the Euclidean distance. In other words, if the cosine is zero, then the images are totally dissimilar.

2.5.6 Effect of the Eyebrows and Upper and Lower Eyelids

The eyelids and eyebrows hide a part of the iris. This affects the performance of the authentication process.

To overcome these problems, Daugman (2004) and Wildes (1997) modeled the upper and lower eyelids with parabolic arcs. Miyazawa et al. (2005) and Daouk et al. (2002) have used only the lower part of the iris for authentication. This method has proven an effectiveness of 100% at the level of verification (Daouk et al. 2002). Tian et al. (2006) adopted the linear Hough transform. They used the model with three lines to approximate the eyelid from each eyelid edge. They also used an adaptive threshold to locate the eyebrows by comparing with the average of the gray level in the region of the iris. Meng and Xu (2006) sliced an upper segment of $\pi/4$ to $3\pi/4$ and a lower segment of $\pi5/4$ to $7\pi/4$ in the region of the iris in order to obtain more information for the process of feature extraction. This method lacks precision as more than half of the information is excluded.

However, Xu et al. (2006) proposed an effective method of eliminating the effects of the eyelids and the eyebrows of the normalized iris image. This method consists in dividing the region of the iris into eight blocks of fixed size (e.g., 9×360). Thus, they have selected eight sub-blocks of fixed size (e.g., 9×45) in each block, so that the variation of the seventh sub-block is less than the threshold. This method has shown a weakness in the models with zero presence of eyelids/eyelashes since the region of the eyelids/eyelashes is considered very large.

2.6 Discrete Geometry in the Definition of an Edge in an Image

We define a location of an edge in a given image by determining a form with specific characteristics (i.e., a circle has a radius, r, and a center, C, of coordinates (x, y)).

We consider that the image is represented as a matrix $A(i, j)$, in which i is the row indexes and j the column index as shown in Figure 2.9. This matrix contains zero and one (Chassery and Montanvert 1991).

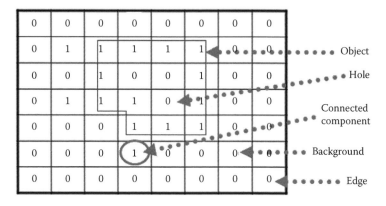

Figure 2.9 Objects, background, hole, and edge in an image.

The elements of value one define the object (e.g., presence of an edge) in the image. The values zero represent the background of the image (i.e., composed of a single connected component[*]) and the holes (i.e., not connected to the edge of the image). The edge of the image has no object point.

2.7 Pretopological Aspects in Image Classification

This section represents the different pretopological aspects that contribute to defining methods of image classification in a database to find them quickly.

2.7.1 Overview of the Pretopological Spaces

Before detailing pretopological aspects in the classification, it is necessary to define the pretopological space (E, i, a)—some notations that will be references to the concepts of pretopology (Belmandt 1993). In the following, we denote:

- E: given set nonempty
- i: interior of E
- a: adherence of E

---

[*] The notion of connectedness in an image is related to the fact that two points are linked together in one image and it is impossible to isolate them.

The interior of part Å on E is defined as the complement of the edge. It is designated by

$$\mathring{A} = i(A) \qquad (2.1)$$

The adherence of a subset A of E, designated by the following, is the smallest closed set of E containing A:

$$\overline{A} = a(A) \qquad (2.2)$$

Note that a closed set is the union of the interior of A and the edge of A. An open set represents the interior of A. Figure 2.10 illustrates these representations.

The notion of idempotence can take several paths. In particular, an application f is idempotent if $f \circ f = f$. In other words, if $f: A \to B$ is idempotent, then if $f(a) = b$, then $f(f(a)) = b$, with a ε A, b ε B and $B \subset A$.

Given E a finite nonempty set, a part Y of $\Pi(E)$ is a prefilter on E if it verifies the stability property by passing any superset

$$\forall \, U \, \varepsilon \, Y, \, \forall \, H \, \varepsilon \, \Pi(E), \, U \subset H \to H \, \varepsilon \, Y \qquad (2.3)$$

where $\Pi(E)$ is the set of nondegenerated (i.e., does not contain $\varnothing$) proper (i.e., $\forall$ A part of E, $A^s = \{B \subset E/B \supset A\}$ then $A^s = \{E\}$) prefilters.

We call the neighborhood of base of an element x belonging to E the database of neighborhood containing a single element associated with x. It is defined by $\varsigma(x)$, where ς is an application of E in $\Pi(E)$, with

$$x \in \varsigma(x), \, \varsigma(x) \subset E \qquad (2.4)$$

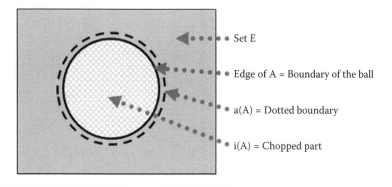

Figure 2.10 Edge, adherence, and interior of an object.

We also designate by neighborhood of x each element "nearest" to x and every element "farthest" from x and not part of the neighborhood of x.

A pretopological space is of type ç if and only if for all part A and part B of E, such that B ⊂ A, we have a(B) ⊂ a(A) and i(B) ⊂ i(A). In addition, for all x belonging to E and for each V element of ç(x), we have

$$ç(x) = \{V \subset E/x \in i(V)\} \tag{2.5}$$

where ç(x) is called the neighborhood of x.

We define a relationship of refinement "finesse" between spaces of type ç by the following: If x is an element of E, ç(x) and ç'(x) (ç(x) ⊂ ç'(x)) are the prefilters of neighborhoods of x, respectively, for the pretopological spaces (E,i,a) and (E,i',a'), then i' is more refined than i if for each subset A of E i(A) ⊂ i'(A). Similarly, i is considered as less refined than i'.

A pretopological space is of type $ç_D$ if and only if for each part A and part B of E we have a(A ∩ B) = a(A) ∩ a(B) and i(A ∪ B) = i(A) ∪ i(B). In addition, for each x element of E, and A and B two elements of ç(x), x is then an element of i(A) and i(B); therefore, i(A) ∪ i(B) i is equal to i(A ∪ B). In this case, we have A ∪ B belongs to ç(x).

A pretopological space is of type $ç_S$ if and only if for every subset A of E we have

$$a(A) = \bigcup_{x \in A} a(\{x\})$$

Moreover, for each x element of E, the intersection G (x) of neighborhoods of x is a neighborhood of x; therefore,

$$G(x) \in ç(x)$$

Note that a pretopological space of type $ç_S$ is of type $ç_D$, but the converse is not true in general.

A binary relation P is defined on a pretopological space of type $ç_S$ by

$$x \, P \, y \text{ if and only if } y \in G(x)$$

where x and y are elements of E.

A binary relation P is called more refined than a binary relation P' if and only if for each x ∈ E, P(x) ⊂ P'(x), or x P' y whenever x P y.

On the other hand, the binary relation P′ is called less refined than binary relation P.

2.7.2 Different Types of Pretopological Structures

Among the pretopological structures (Belmandt 1993) we can mention the following:

- Preuniform structures
- Induced structures
- Poor structures

In the remaining, "s" designates a pretopological structure on a non-empty set E.

We call a preuniform structure on a set E, the structure constituted by the data of a family Y of parts ExE such as

1. Y is a prefilter of part ExE.
2. $\forall$ U, U ε Y we have $\Delta_E \subset$ U, with Δ_E diagonal of ExE.

An induced structure $s_A = (i_A, a_A)$, A a subset of E, verifies the following conditions:

1. If s is of type ς_D, s_A is of type ς_D.
2. If s is of type ς_S, s_A is of type ς_S.
3. If s is idempotent, s_A is idempotent.
4. In any case, if K is a close (respectively, open) of E for s, then $K \cup A$ is a close part (respectively, open) of A for s_A.
5. If s is of type ς_D and if A is open for s, then
   - any part G of A open for s_A is open in E for s
   - any part K of A close for s_A, $K \cap (E-A)$ is close of E for s
6. If A is a close of E for s, if K is a close of A for s_A, then K is also a close of E for s.

A poor structure s = (i,a) is free, under certain circumstances, of the following constraints applied to any subset A of a set E:

- For i, the interior, $i(A) \subset A$ and $i(E) = E$.
- For a, the adherence, $a(A) \supset A$ and $a(\emptyset) = \emptyset$.

This poor structure uses certain specific pretopological structures of type ς, ς_D, or others.

2.7.3 Classification Method

The classification consists in grouping similar elements of E (E ≠ ∅). Each group represents a nonempty and homogeneous subset. All these groups cover the set E.

We can mention two main classification methods (Saporta 1990):

1. The hierarchical classification leading to merge in the same group two elements of E that have a given level of accuracy, while at a higher level of accuracy, they will be distinguished and belong to two different subgroups
2. The nonhierarchical or partitioning classification, leading to decomposition of the set E of all elements into m disjoint sets or equivalence classes whose number of classes m is fixed

A hierarchy H is called indexed if there is a function ind: $H \rightarrow \mathbb{R}^+$ such as

- $\forall\ H\ \varepsilon\ H,\ \forall\ H'\ \varepsilon\ H$, then if $H' \subset H \Rightarrow ind(H') < ind(H)$
- $\forall\ x\ \varepsilon\ E$, then $ind(\{x\}) = 0$ and $ind(E) \leq 1$

For hierarchical classification methods, we can distinguish two concepts:

1. The concept of dissimilarity on E, representing the entire application d:
   - $ExE \rightarrow \mathbb{R}^+$ that verifies, for all $x \in E$ and for all $y \in E$:
     - $d(x, y) = d(y, x)$
     - $d(x, y) = 0 \Leftrightarrow x = y$
2. The concept of similarity on E, representing any application s:
   - $ExE \rightarrow \mathbb{R}^+$ which verifies, for all $x \in E$ and for all $y \in E$:
     - $s(x, y) = s(y, x)$
     - $s(x, x) \geq s(x, y)$

Among the methods of constructing a hierarchy indexed by trees (Azzag, Guinot, and Venturini 2004), we detail the method of construction of a B-tree (Mannino 2004). A B-tree is a balanced tree. Such a tree is implemented in the management mechanisms of databases and file systems. It stores data in a sorted way and allows execution of operations of insertions and deletions in logarithmic amortized time.

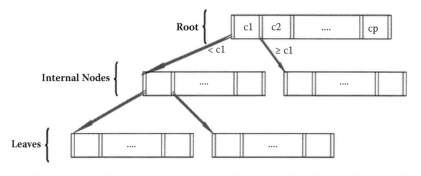

Figure 2.11 General representation of a B-tree.

The structure of a B-tree is represented by nodes of n-levels (Mannino 2004) as shown in Figure 2.11. Level 1 represents the root of the tree, the last level represents the leaves, and other levels (between the root and leaves) represent the internal nodes. Each node contains keys and pointers. The left pointer of a key points to values that are strictly less than that of this key, and the right pointer points to values that are greater than or equal to the value of this key. The general principle is to allow nodes to have more than one key. This minimizes the size of the tree and reduces the number of balancing operations. In addition, the B-tree grows from the root and in a dynamic way.

Each L-U tree B (L and U, two non-zero natural integers such as L ≤ U) is defined as the following:

- Each node except the root has at least L-1 keys, at most U-1 keys, and at most U children.
- For each internal node, the number of children is equal to the number of keys incremented by one unit.

For a tree L-U, we have n nodes and n children with L ≤ n ≤ U.

The research in a B-tree (Mannino 2004) is recursive. Starting from the root to each node, we choose the child subtree whose keys are between the same limits as those of the desired key.

The insertion requires searching the node where the new key should be inserted (Mannino 2004). This research is conducted recursively, from the leaf level by going up to the root, so that a node has too many keys or not. If this node has an acceptable number of keys, we insert this key without changing the structure of the tree. Otherwise, we split it into two nodes, each with a minimum number of keys, and then we go up the middle key, which is then inserted into the parent node. This last ends up with too many children. The process continues until the root is reached. In this case, we go up the middle key in a new root, which will generate as child nodes the two nodes created from the old root. The process of division of the root may happen if U ≥ 2L; otherwise, the new nodes will not possess enough keys.

The deletion of a key proceeds by finding that value in the node that contains it and then its deletion (Mannino 2004). The process of deletion (Mannino 2004) is described as follows:

- If this node is an internal node, we perform the search for a key k left-most in the right subtree to delete the key or right-most in the left subtree. This key k belongs to a leaf. We can swap it with the key to remove later. As k belongs to a leaf, we will pass to the next case.
- If this node is a leaf, or it still has enough keys and the algorithm terminates, or it has less than L-1 keys, then we are in one of the two following situations:
  1. A brother to the right or left key has enough power to pull a leaf in question; in this case, this key overrides the key that separates the two subtrees in the tree father, which is itself in the leaf in question.
  2. Not any brother has enough keys; in this case, the father passed one of the keys in one of two (or only) brothers to allow the leaf to merge with it. This can, however, lead to the father not having enough keys. We reiterate the algorithm: If the node has a brother with enough keys, the nearest key will be exchanged with the key of the father, and then the key of the father and his new offspring are returned to the node that needs a key, if not we do a fusion with a key of the father and so on. If we get to the root and it has less than L elements, we merge both children to give a new root.

2.8 Conclusion

In this chapter we introduced biometric methods capable of providing an authentication level of access valid to systems including data.

We made a comparative study between these different methods to show that the method of biometric authentication based on the iris of the eye is the most powerful due to its precision and minimal error rates. Thus, we have presented various algorithms of recognition of the iris of the eye stating, at each phase of each algorithm, the methods adopted. This study helped us to develop an algorithm providing accurate authentication with less time. Aspects of discrete geometry are shown to define the concept of the localization of edges of the iris in an image. Similarly, pretopological aspects on the classification of objects are detailed to present the classification method of iris in a database.

The advantages of the biometric method using the iris of the eye justify our choice. To enhance the security of data access, we have integrated it with a cryptographic method that will be detailed in the following chapter.

3

CRYPTOGRAPHY

In this chapter we present a description and a comparison of different methods of cryptography: symmetric cryptography, asymmetric cryptography, digital signature, and digital certificate. The comparative study of these methods helped us decide to use the method based on asymmetric cryptography using the algorithm Rivest–Shamir–Adleman (RSA) (Stallings 1999). This method has a mode of transmission faster than other algorithms that can be used (MD4, MD5, SHA, and SHA-1).

3.1 Introduction

Cryptography is a method to make information unreadable to ensure access to a single authenticated recipient (Otto 2004). This approach ensures the anonymity and the security for confidentiality, authenticity, and integrity of information (Futura-Sciences 2008).

The conversion of data is performed by means of a key. A key is information known only to the sender and the receiver and serves to control the processes of encryption and decryption (Microsoft 2002). It is similar to a password and transmitted separately via a secure channel. It must be complex enough not to be violated. Several cryptographic techniques have been developed to secure access to data. These techniques are detailed in the next section.

3.2 Different Cryptographic Methods

Four methods of data encryption have been proposed for transmission security: symmetric cryptography, asymmetric cryptography, digital signature, and digital certificate.

3.2.1 Symmetric Cryptography

This method is known as single key cryptography (Microsoft 2002). It uses the same key for encryption and decryption using various algorithms. Among these algorithms we have the data encryption security (DES) algorithm of 56 bits, the DES algorithm with three keys (triple DES) of 56 bits each and a total of 168 bits, and the advanced encryption standard (AES) algorithm of 128 bits (Stallings 1999). This technique was adopted in 1977 by the National Institute of Standards and Technology (NIST).

3.2.1.1 Overview of the Method   Symmetric cryptography, shown in Figure 3.1, works as follows:

- The sender, Alice, wanted to send to the receiver, Bob, in encrypted form, a clear message: $X = [X_1, X_2, \ldots, X_M]$. The M elements of X belong to a specific alphabetic system composed traditionally of 26 uppercase alphabetic letters and binary alphabets {0,1}.
- A key, $K = K = [K_1, K_2, \ldots, K_J]$, is generated. This key can be generated and transmitted by Alice to Bob over a secure channel or supplied by a third party who will pass it in a secure way to Alice and Bob.
- Alice encrypts the clear message X to a ciphertext $Y = [Y_1, Y_2, \ldots, Y_N]$ via an encryption algorithm, using the key K. The ciphertext based on the clear message is represented by $Y = E_K(X)$, where E_K is the encryption method.

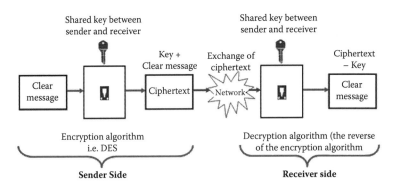

Figure 3.1   Model of symmetric encryption.

- Alice sends the ciphertext Y to Bob. This last one decrypts, using the key K, the ciphertext to clear message. The clear message, depending on the ciphertext, is represented by $X = D_K(Y)$, where D_K is the method of deciphering.

3.2.1.2 Existing Methods Among the methods of symmetric cryptography, we can list the Caesar cipher and the method of disposable masks.

- The Caesar cipher is a simple substitution method. It is used to replace each letter in the message by the one that follows in the alphabet (or two or more later) (Stallings 1999). The Caesar encryption algorithm is expressed by $C = E(p) = (p+k)$ mod (26), where E is the encoding function. Each letter p of the clear message is replaced by a letter C in the ciphertext. K represents the offset value applied to each letter of the alphabet and can have a value from 1 to 25. In addition, the clear message is presented in lowercase letters, while the ciphertext is presented in capital letters. The decryption algorithm is represented by $p = D(C) = (C-k)$ mod (26), where D is the decoding function.
- The method of disposable masks is a method of symmetric cryptography where substitution ciphers have key length equal to the text (Florin and Natkin 2003). This method is based on a longer key, more complex, and that the offset of each letter is not constant. This method hides the original meaning of the text and holds up better than the first attack. However, it is difficult to decrypt the ciphertext in case of data loss.

3.2.1.3 Fields of Application This technique is applied in several fields, such as wireless sensor networks to secure routing protocols (Chen, Zhang, and Hu 2008), the Internet to configure Internet protocol (IP) addresses to protect intellectual property rights against any attack (Güneysu, Möller, and Paar 2007), and the medical sector to secure medical images and their transmission (Ashtiyani, Birgani, and Hosseini 2008).

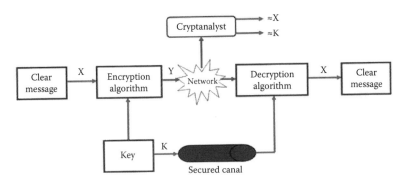

Figure 3.2   Model of encryption system.

3.2.1.4 Advantages and Disadvantages   This method has excellent performance in speed of encryption and decryption. However, the delivery of the key to the receiver makes a problem, since there is a risk of stealing of the key by an unauthorized person.

Moreover, there are two approaches to this method of attack (Stallings 1999), cryptanalysis and brute-force attack (Figure 3.2):

1. The cryptanalysis is based on the nature of the algorithm and sometimes on some knowledge of general characteristics of the clear message or the ciphertext.
2. In the brute-force attack, the attacker tries every possible key to the ciphertext until it leads to a clear message.

In addition to the risks of attack, the key length is a very important factor. The longer the key is, the lower is the risk of attack, but the transmission time is long.

3.2.2 Asymmetric Cryptography

Asymmetric cryptography is known as public key cryptography or double key. It was proposed by the National Security Agency (NSA) in 1960. Then, Whitfield Diffie and Martin Hellman (Stalling 1999) applied the concept of public key cryptography in 1976 at Stanford University. This method uses the cryptographic algorithm Rivest–Shamir–Adleman (RSA) with a key of 128 bits (Stallings 1999). This algorithm was proposed by Ron Rivest, Adi Shamir, and Len Adleman in 1977 and published in 1978. In the 1990s, two hashing algorithms in one-way Message Digest version 4 (MD4) and

Message Digest version 5 (MD5) were published by Ron Rivest. Two other hash algorithms, secure hash algorithm (SHA) and secure hash algorithm 1 (SHA-1), were proposed by NIST. These algorithms are complex and have a long key. This leads to slow transmission and a high level of security, especially against the brute-force attacks.

3.2.2.1 Overview of the Method Asymmetric cryptography uses two keys: a public key to encrypt data and a private key to decrypt them. Each user generates a pair of keys. The public key is stored in a public register accessible to everyone. The private key is kept locally at the user and in secret. Each participant may have a collection of public keys of others.

This method, shown in Figure 3.3, works as follows (Stallings 1999):

- The sender, Alice, wanted to send to a receiver, Bob, a clear message: $X = [X_1, X_2, \ldots, X_M]$. The M elements of X belong to the set of alphabetic letters.
- Bob generates the key pair: public key KU_b and private key KR_b. KR_b is known only by Bob, but KU_b is accessible to all the public and especially to Alice.
- Alice encrypts the clear message using Bob's public key KU_b to obtain a ciphertext $Y = [Y_1, Y_2, \ldots, Y_N]$. Text Y, function of X, is represented by $Y = E_{KUb}(X)$, where E_{KUb} is the encoding function.
- Alice sends the ciphertext Y to Bob.
- When B receives the ciphertext, he decrypts it using its own private key KR_b to obtain the clear message $X = D_{KRb}(Y)$, where D_{KRb} is the decoding function.

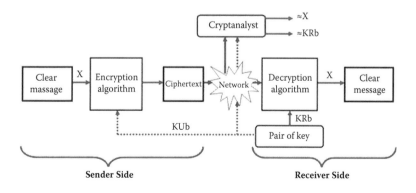

Figure 3.3 Model of asymmetric encryption.

3.2.2.2 Fields of Application Asymmetric cryptography has been adopted in several areas, such as mobile networks for the safe delivery of information and communication between different users (Grecas, Maniatis, and Venieris 2001; Capkun, Hubaux, and Buttuán 2006), Internet networks to protect users' applications against unauthorized interference (Parnes, Synnes, and Schefström 1999), and wireless networks to secure the transmission of data (Lee et al. 2008).

3.2.2.3 Advantages and Disadvantages The fact that the method of asymmetric cryptography does not distribute the private key protects transmitted information against unauthorized access.

The processes of encryption and decryption are executed quickly. On the other hand, an attacker who knows the algorithms of encryption (E) and decryption (D) can easily access data. If the attacker knows only the ciphertext Y and the public key KU_b of Bob, without having access to the private key of Bob, KR_b, or to the clear message X, then it can only try to generate by successive approximations a clear message close to X and/or a private key close to KR_b.

3.2.3 Digital Signature

The digital signature provides authentication and data integrity.

3.2.3.1 Overview of the Method The digital signature method, shown in Figure 3.4, operates as follows (Stallings 1999):

- The sender, Alice, wanted to send a clear message X, in encrypted form, to the receiver, Bob. Alice prepares the clear message X.
- Alice encrypts the message X using its private key KRa to get the ciphertext Y.
- Bob decrypts the ciphertext using the public key of Alice, KUa.

3.2.3.2 Fields of Application This method is used in several areas, such as the commercial sector (e.g., e-commerce), to protect commercial transactions (Chen, Richard, and Chen 2003); the specialized networks in the transmission of the authentication messages of remote associates; and adopting smart cards to protect their privacy (Berta and Vajda 2003).

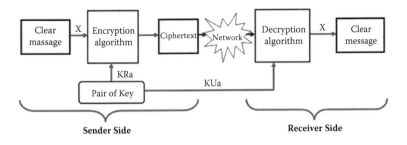

Figure 3.4 Model of digital signature.

3.2.3.3 Advantages and Disadvantages This method requires much time, but it is more secure than the method of asymmetric cryptography. It has shown great efficiency in business operations (Chen et al. 2003). However, the possibility for the user to create its digital signature in places where a security key is not guaranteed makes for a risk of attacks (Campbell 2003), since the process of creation of this signature requests the private key of the given person.

3.2.4 Digital Certificate

The digital certificate is an electronic license issued by service agency certification authorities (CAs) such as VeriSign or GlobalSign (Microsoft 2002). This agency is a company whose operation is based on trust. It establishes and verifies the authenticity of certificates issued to users or other authority certificates. The digital certificate is organized as a hierarchy of relational parent/child, as shown in Figure 3.5.

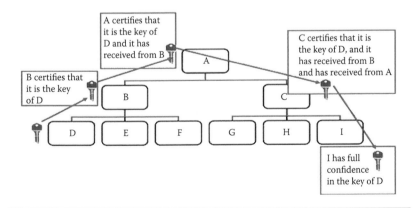

Figure 3.5 Model of digital certificate.

3.2.4.1 Overview of the Method   This method works as follows:

- The user A certifies the identity of the user B.
- A authorizes B to certify the identity of users D, E, and F and generates the key of B.
- B certifies the identity of D, E, and F and generates their keys.
- A certifies the identity of the user C.
- A authorizes C to certify the identity of users G, H, and I and generates the key of C.
- C confirms the identity of users G, H, and I and generates their keys.

The digital certificate contains the identity of the owner certificate, the identity of the CA that issued the certificate, the authentication extensions, the value of the owner's public key, the validation date of the certificate, and the digital signature of the CA to ensure data integrity.

3.2.4.2 Fields of Application   The digital certificate is used to certify the online identities of individuals, organizations, or computers (Stallings 1999). It is also integrated into the authentication process for network access, such as commercial networks (e.g., e-commerce), financial networks (e.g., e-banking), and networks on interpersonal relationships (Data Investment Consult 1999, 2000; Reuters 2006).

3.2.4.3 Advantages and Disadvantages   This method is easy to use to manage the security of public and private networks in a hierarchical form (Microsoft 2002). It personalizes the application of cryptographic algorithms (Markovie 2007). It also accepts users with a pair of keys, two pairs of keys, or a combination system. However, this method has a lack of confidence in case of a nonhierarchical structure. In this case, it is difficult to validate certificates issued by anonymous CAs, since there is no single CA route (Zhou and Harn 2008).

3.3 Recapitulative Table of Comparison of Different Cryptographic Methods

Table 3.1 compares the different cryptographic methods by showing the differences as well as the advantages and the disadvantages of each (Stallings 1999; Florin and Natkin 2003).

Table 3.1 Comparison of Different Cryptographic Methods

	SYMMETRIC CRYPTOGRAPHY	ASYMMETRIC CRYPTOGRAPHY	DIGITAL SIGNATURE	DIGITAL CERTIFICATE
Key	Unique secret key shared between sender and receiver	A pair of keys: private secret key to its owner and public key accessible to all public	A pair of keys: private secret key to its owner and public key accessible to all public	An electronic license
Risk of attack	Easy to the clear message and the ciphertext	Encrypted text, but not easy	Private key	Ciphertext
Level of security	Risky	Effective	Effective	Depends on trust
Execution time	Fast	Slow	Very slow	Varies depending on the structure

3.4 Conclusion

In this chapter we have detailed the various methods of cryptography: symmetric cryptography, asymmetric cryptography, digital signature, and digital certificate. These methods each have strengths and weaknesses as shown in Table 3.1. This comparative study justifies our choice for the method of asymmetric cryptography.

We consider that the asymmetric cryptography applied on the biometric method of the iris of the eye enhances the security of access to data. In the advance applications, the notion of agent and multiagent system (MAS) is integrated to treat the complex problems of security in the computer networks. The next chapter will take this approach.

4

MULTIAGENT SYSTEM

Agents and multiagent systems, inspired by the field of artificial intelligence, have demonstrated their effectiveness in solving complex problems in various fields. We assumed to use them to manage operations in a well-organized and coherent way. Specialized agents will be integrated into a network in order to control certain parameters and solve specific problems in terms of security, confidentiality, and data integrity. In this chapter we present the different types of agents and their properties depending on the nature of the environment. We detail the communication, the cooperation, and the architectures of agents.

4.1 Introduction

A multiagent system (MAS) is a set of agents interacting in an environment in order to achieve the realization of a global objective (Meskaoui 2005). This interaction is expressed by a communication or cooperation and coordination between agents. A multiagent system is characterized by the number of agents and their interactions, the mechanisms and the types of communication, the behavior, and the organization and the control of each agent or the representation of the environment. This set of interactions from the behavior of simple small agents has resulted in the ability to solve problems of great complexity.

In the field of artificial intelligence, an agent is an intelligent entity acting in a rational and intentional way to its goals and the current state of knowledge (Demazeau and Müller 1991). These intelligent agents constitute the first category of tools (Pujolle 2003), where the widespread introduction could modify the environment of management and control by making them more autonomous and more reactive. They are classified into two categories (Ferber 1997):

1. A *physical entity* is something that acts in the real world, such as a robot, an airplane, or a car.
2. A *virtual entity* is something that does not physically exist, such as a software component or a module.

These agents are applied in several areas, such as

- Data transmission in an estimated mobile environment (Nam and Park 2001)
- Dynamic development of robotic soccer (Bo and Qinghua 2000)
- Virtual shopping on the Internet (Yu, Wu, and Wu 2004)
- Modeling needs of the body and the brain (Yu, Wu, and Hou 2004)
- Systems of social rights (Boella and van der Torre 2005)
- The borrowing process (Huhns and Singh 1998)
- Application of automatic recording systems of e-mail accounts (Cabri, Ferrari, and Leonardi 2003)
- Simulation of distributed data (Wilson et al. 2001)
- Communication services of administration (Bawany, Paracha, and Naz 2004)
- Electronic products (Kearney 1996)
- Care for elderly and handicapped people (Ishikawa et al. 2000)
- Electronic commerce (Kang, Park, and Koo 2003).

4.2 Properties of Agents

The agents are characterized by their properties that determine their capabilities (Ferber 1997). Among these properties, we mention the following:

- *Autonomy:* The agent is able to take the decision to change its behavior with respect to changes in its environment without waiting for human interventions and independently of another agent in the same environment.
- *Flexibility:* The agent is able to change its behavior dynamically in order to adapt to various changes in its environment.
- *Adaptability:* The ability of the agent to adapt its needs and behaviors to the availability of resources in its environment.

- *Collaboration and cooperation of tasks:* The agent is able to cooperate with other agents in its own field in order to achieve a common goal and to share knowledge and learning experiences.
- *Mobility:* The agent is able to move from one host to another over the network in order to perform some assigned duties. This property is not required for all agents.

4.3 Types of Agents

Several types of agents (Ferber 1997) have been defined as cognitive, reactive, hybrid, and mobile.

- The cognitive agents are the most represented in the field of distributed artificial intelligence (DAI). They have an explicit representation of the environment and other agents. They take into account their past and operate in a social mode of organization. Their number is very small in the systems that have adopted this type of agent. Most cognitive agents are intentional. They use the concepts of intention, commitment, and explicit partial plans allowing them to reach their goals. These agents coordinate their activities and sometimes have to negotiate among themselves to resolve conflict. These agents may also be of the "cognitive reflex" or "module" type. They cooperate with each other and involve concepts of competencies, mutual representations, and task allocation. They act according to regulations and social legislation—hence, the aspect of *organized agents.* These agents communicate with each other in an intentional way using a language. They use communication protocols by sharing of information when the solution of problems is centralized in a global data structure and shared by all agents. Otherwise, they use communication protocols for sending messages characterized by the total distribution of knowledge, partial results, and the methods used to achieve a result. Several types of complexity can be considered as processes in which actors are implementing communication primitives and communicating modules that use specialized communication protocols (e.g., requests or commands). As well, there are calculating agents that directly respond to queries addressed to them.

- The reactive agents are not an explicit representation of their environment and cannot take into account their past. Their modes of operation are simple and wired, of stimulus–response type. Their numbers are very high in systems that have adopted this type of agent. These agents communicate with each other in nonintentional ways and leave their presence signals or signs that may be perceived by other agents. Several types of complexity can be considered (Ferber 1997):
  - Stimulus–response levels that are simple reactions to events
  - Levels of coordination of elementary actions that are mechanisms of inhibition and the relationship between elementary actions
  - Levels of reactive cooperation that are recruitment mechanisms between agents and an aggregation of elementary agents
  - Levels of reproduction, which are mechanisms of reproduction of reactive agents and organizational levels of reactive agents
- The hybrid agents consist of a combination of two or more philosophies of agents in a single agent. These philosophies include mobile philosophy, the philosophy of the reactive agent, and the philosophy of the cognitive agent. The key assumption for agents or hybrid architectures, for some applications, is the belief that the gains in having a combination of philosophies in a single agent are larger than those obtained from the same agent based on one philosophy. However, it is necessary to prevent a conflict between the various philosophies by well-defined interfaces between them (Chatley 1997).
- Mobile agents are computational software processes. They are able to surf remote networks, interacting with foreign hosts, and to collect information to perform specific tasks requested by their owner.

4.4 Communication of Agents

Communications in multiagent systems are the basis of interaction and social organization (Ferber 1997). The communication between agents allows them to cooperate, coordinate their actions, perform tasks in common, and become genuine social beings. The communication is described as a form of interaction in which the dynamic relationship between agents is expressed through mediators—signals that, when interpreted, will have effects on these agents.

Several theories of communication have been proposed, but they are all based on variants of the theory of communication, which is the result of research of 40 years of telecommunications developed by Shannon and Weaver (1948). In this model, the act of communication consists of transmitting information from a sender to a receiver. This information is encoded by the sender and decoded by the receiver. It is transmitted in a channel (or medium). The context is the situation in which the speakers are placed.

The communication is not just a simple transmission of information or exchange of messages, but it takes more elaborate forms such as language acts and conversational structures that focus on the notion of interaction in communications (Ferber 1997). The communication is then considered as an ongoing social process incorporating multiple modes of behavior such as speech, gesture, look, facial expression, and interpersonal space (Ferber 1997).

4.4.1 Acts of Language

The acts of language represent a major theory of the philosophy of language, which represents an important interest for the analysis of point-to-point symbolic communications in multiagent systems (Ferber 1997). These acts point at the set of intentional actions achieved in the course of a communication. There are several types of acts of language (Ferber 1997):

- The assertive serves to provide information about the world by saying something like "it is lovely weather" or "herbivores do not eat meat."
- The directive is used to give instructions to the recipient, such as "give me the table" or "what is the third letter in the alphabet?"
- The promissive commits the speaker to perform certain acts in the future, such as "I promise to send you the document tomorrow."

- The expressive serves to provide the recipient with details of the speaker's mental state such as "I am happy," "I apologize for yesterday," or "thank you."
- The declarative performs an act thereby to utter the statement such as "I curse you" or "I sentence you to 2 years in prison."

4.4.2 Acts of Conversation

Conversation is described as a sequence of states, linked by transitions, representing communications between agents. The modeling of conversations requires the definition of protocols (i.e., valid sequences of messages). These protocols are modeled as a finite state machine or Petri network (Ferber 1997). The model of a finite state machine is represented as a sequence of states linked by transitions (Winograd and Flores 1986). This is to clarify the structure of conversations when they occur in isolation—in other words, when the conversation is reduced to a single process.

The model of Petri networks is adopted when agents are engaged in several conversations at once, and it is necessary to manage these various conversations. These networks are used to model protocols in distributed systems (Estraillier and Girault 1992). As well, they model multiple simultaneous communications with multiple parties. In this case, the message is numbered according to a serial number corresponding to the conversation to which it belongs. This is done in order to avoid confusion between messages of different conversations.

4.4.3 Languages of Communication

Agents have the ability to communicate and cooperate to achieve their goal. In this context, it is necessary to have a structured language for communicating agents in a different group. This language of interaction between agents should have semantics using the recurrence of acts of language (Chaib-Draa and Vanderveken 1999) and a framework for interpreting acts of conversation (Vongkasem and Chaib-Draa 1999). This language introduces standard message types to allow agents to interpret identically and understand the content in the same way. Mayfield, Labrou, and Finin (1995) identified the need for defining languages of agents. These needs are divided into seven categories: form, content, semantics, implementation, network, environmental, and reliability. At some point, these needs may be in conflict with each other. Therefore, the person responsible works to balance these various needs in order to avoid conflict.

Several studies have been conducted to define standardized languages to support interagent communication. Among these languages we mention "knowledge query manipulation language" (KQML) and "agent communication language" (ACL).

- The KQML language was proposed to support communication between software agents (Labrou and Finin 1994). This language defines a set of message types called "performative" and rules that define the behavior suggested to the agents who receive these messages (Finin et al. 1994). This language is based on the theory of "speech act," which defines the acts of communication. The communication is a way of expressing a certain attitude. Thus, the type of speech act is the type of expressed attitude. For example, an apology expresses regret, while a request expresses a desire. As an act of communication, the speech act succeeds if the audience identifies the attitude expressed in accordance with the intention of the speaker. The KQML message types are diverse in nature (Finin et al. 1995): routing instructions of information (e.g., "forward" and "broadcast"), simple queries and assertions (e.g., "ask," "tell"), persistent commands ("subscribe," "monitor"), and commands that allow consumer agents to ask intermediate agents to find relevant supplier agents ("advertise," "recommend," "recruit," and "broker").

- The ACL language is semantically richer than KQML language. This language is proposed by the Foundation for Intelligent Physical Agents (FIPA), which handles the standardization of communications between agents (Labrou and Finin 1997). It is based on language theory. The language ACL took advantage of the search results of KQML. These two languages come closer together in acts of language, but not at the level of semantics. The ACL language is treated more accurately.

4.5 Cooperation of Agents

A multiagent system differs from a collection of independent agents. These agents have several actions between them to achieve specific goals (Cite-Sciences 2005). Among these actions we can mention interaction and cooperation, coordination, and negotiation.

4.5.1 Interactions

Several agents can interact together to achieve a task or a particular goal together (Ferber 1997). These agents communicate with each other directly, through another agent, or even acting on their environment.

The interactions of agents in multiagent systems are driven by the interdependence of these agents according to three criteria:

1. Goals (or intentions) of agents, whether their goals can be compatible or not
2. Relations that agents maintain toward the resources they have—in other words, agents may want resources that others have
3. Means (or competence) available to them to achieve their ends—in other words, an agent A can have a capacity necessary for an agent B for the fulfillment of one of B's plans of action

The cooperation in a group of agents designed to increase the number of tasks executed in parallel with the parallelism and the achievable tasks by sharing resources increases the chances of completing tasks by duplicating and using possibly different modes of realization, as well as reducing interference between tasks by avoiding negative interactions (Durfee and Lesser 1989).

Cooperation is implemented, for example, in approaches to distributed cooperative problem solving (CDPS). In these approaches, agents cooperate with each other to solve problems they cannot solve individually. Other views on cooperation have been developed by several researchers. Ferber (1997) considered cooperation as an attitude adopted by agents that decide to work together. Otherwise, it was considered an outsider in a multiagent system that interprets the behavior of agents according to predetermined criteria such as independence of the actions or the number of communications made.

Galliers (1988) and Conte, Miceli, and Castelfranchi (1991) have considered cooperation an essential element of social activity.

4.5.2 Coordination

Different points of view exist on the coordination of actions among a group of agents. Malone (1990) specifies two fundamental components of coordination between agents: the communication of intermediate results and the allocation of scarce resources. In the case of communication of intermediate results, the agents are able to communicate with each other so that they can share intermediate results. For the allocation of shared resources, agents are able to make transfers of resources. This imposes certain behaviors on particular agents.

Three fundamental processes of coordination are identified by Mintzberg (1979):

1. Mutual adjustment is the form of the simplest coordination that occurs when two or more agents agree to share resources to achieve a common goal. In this case, the agents exchange many pieces of information and make several adjustments to their own behavior, taking into account behaviors of other agents. In this form of coordination, no agent has control over other agents and the joint decision-making process is such as that in markets.

2. Direct supervision is the fact that one or more agents have already established a relationship in which one agent has control over the others. This relationship is established by mutual adjustment: For example, an employee agrees to follow the instructions of the supervisor. In this form of coordination, the agent supervisor monitors the use of shared resources by subordinate agents, such as human resources or computation time. This agent supervisor may also impose certain behaviors.

3. Coordination by standardization is integrated in business and in computer systems. In this form of coordination, the supervisor coordinates the activities by establishing procedures to be followed by subordinates in identified situations.

According to Conte et al. (1991), coordination is a key issue for multiagent systems and for the resolution of distributed systems. Indeed, the coordination that provides consistent behavior of the group of agents has a centralizing agent that holds high-level information about other agents. Thus, the centralizing agent creates action plans and assigns tasks to various agents of the group. This approach is not practically applicable in real cases since it is difficult to realize such a centralizing agent that can take into account goals, knowledge, and activities of each agent. In this context, the communication load would be enormous and the role of a multiagent system composed of autonomous agents would be lost. We can see that the control and information should be distributed among agents.

Martial (1990) identified two categories of relations between the actions performed simultaneously by multiple agents: negative relationships and positive relationships. Negative (or conflicting) relationships are those that interfere with or prevent a number of actions being carried out simultaneously. They are due to incompatibility of goals or resource conflicts. For example, in a clothing store the two agents, A and B, want to buy one and the same dress. Positive (or synergistic) relationships are those that allow actions to benefit each other. For example, the three agents (A, B, and C) are in a study room with the door open. Agent A feels hot but agent B is bothered by the noise outside; agent C closes the door and operates the air conditioning.

Ferber (1997) considered the coordination of actions in a distributed multiagent system as a fulfillment of tasks by a group of autonomous agents who pursue their own goals. This coordination is involved in defining the sequence of actions to take to avoid the conflict.

4.5.3 Negotiation

Negotiation plays a fundamental role in the cooperation activities by allowing people to resolve conflicts that could put cooperative behavior at risk (Ferber 1997). Durfee and Lesser (1989) define negotiation as the process of improving the agreements and reducing inconsistencies and uncertainty on common points of view or plans of action through the structured exchange of relevant information.

Research in negotiation can be divided into three categories (Müller 1996):

1. The language of negotiation, its semantics, and its use in all protocols
2. Negotiation decisions that are interested in algorithms to compare the subjects of negotiation, the utility functions, and the characterization of agents' preferences
3. The negotiation process that studies general models of negotiation

Several protocols have been adopted for negotiation between agents to solve problems of conflict. Among these protocols we can mention the protocol of the network "contract-net" and the protocol of the cooperation strategy:

- The protocol of the network contract-net is the most used in multiagent systems (Smith and Davis 1980). Agents coordinate their activities through the establishment of contracts to achieve specific goals. In this context, an agent acting as a manager breaks the contract (e.g., a task or problem) into subcontracts, which may be treated by agents of potential contractors. The manager announces each subcontract on a network of agents. The agents receive and evaluate the announcement. The agents have appropriate resources, expertise, or the required information to send to the manager of submission "bids" indicating their ability to perform the task advertised. The manager evaluates the submissions and gives tasks to the most appropriate agents. These agents are called contractors. Finally, managers and contractors exchange the information necessary for the accomplishment of tasks. This protocol is adopted to develop a system of production control (Parunak 1996).
- The protocol of the cooperation strategy was proposed by Cammarata, McArthur, and Steeb (1983), who studied the strategies of cooperation to resolve conflicts between plans of a set of agents. These strategies were applied to the field of air traffic control. This allows each agent (e.g., representing a plane) to build a flight plan to keep a safe distance from other aircraft and satisfy constraints such as reaching the desired destination with minimum fuel consumption. The chosen strategy is called the centralized task. It allows agents involved in a potential conflict situation (e.g., planes that get too close)

to choose one of them to resolve the conflict. This agent acts as a central planner and develops a multiagent plan that specifies the concurrent actions of all aircraft involved. The agents use the negotiation to determine which is most constrained.

4.6 Planning in a Multiagent Environment

Planning is the subfield of DAI that seeks to answer the question: "What should we do?" or "What action should be raised and in which order?" (Allen, Hendler, and Tate 1990; Ferber 1997). In other words, planning is looking for a set of action plans that allows for an initial situation to reach a final situation goal corresponding to the solution of a problem (Portejoie 1991).

Solving a problem consists of two independent phases:

1. Planning
2. Implementation of the plan—the actual passage of the initial state to the final state corresponding to the goal

We call planner the software tool to produce an action plan (Portejoie 1991).

Planning plays an important role in a multiagent system in which several officers belonging to the same environment are involved (Portejoie 1991). These agents help to accomplish a common task or work autonomously.

The interdependence between the actions increases the difficulty of obtaining an order suitable for applying operations (Ferber 1997). Moreover, with the introduction of several agents, two approaches were used to carry out the actions: parallel and sequential. However, these two approaches may contradict. It is in this context that there is interest in planning to build a global plan that will be respected by all agents or allow each to establish its own plan to consider the behavior of others (Portejoie 1991). In this case, the agents are involved in a hierarchical system.

The sequential approach is in the system Stanford Research Institute problem solver (STRIPS) (Ferber 1997). This system is the first planner developed. It is limited to single-agent planning. It is based on state space modeling. The system STRIPS contains the following components (Portejoie 1991):

- A database describes the state of facts (facts base) currently in the world.
- A symbolic model of the environment of the agent is typically represented in a limited subset of first order of the logic.
- There is a symbolic specification of actions that the agent can perform. This specification is usually represented as <precondition, action (or list of additions), effect (or list of withdrawals)>. After application of these actions, the state is changed according to lists of additions and withdrawals that characterize the effects of the modeled actions; one of these two lists is possibly empty.
- A planning algorithm is able to manipulate the defined symbols. It also generates, as output, a plan representing the actions that the agent should do to achieve its goal.

Under the multiagent system, the system STRIPS presents many difficulties in the simultaneous execution of agent actions and interactions that arise from agents (Ferber 1997). Despite these difficulties, several authors have used this system in their planning.

Other approaches of implementation of the centralized multiagent planning have been proposed such as the one of Georgeff (1983). In this approach, the agents' plans are initially created individually. A centralizing agent collects these plans and analyzes them to identify conflicts. This agent tries to resolve conflicts by modifying local plans of other agents and introducing communication commands so that the agents are synchronized in an appropriate way.

However, in an approach of distributed planning, activities are divided into a group of agents, where no agent has a particular control on the other (Portejoie 1991). This raises a certain number of problems that are classified as follows:

- Problems of cohabitation: sharing common resources
- Problems of cooperation: exchange of services
- Problems of communication: exchange of knowledge

Portejoie (1991) noted that "in the context of distributed planning, baseline data should be enriched in order to provide each stakeholder knowledge on background knowledge and knowledge of others," which aims to solve problems based on knowledge.

Durfee and Lesser (1989) proposed an approach called partial global planning (PGP). In this approach, agents build and share plans to identify potential improvements to their coordination.

4.7 Architectures of Agents

Agents and multiagent systems are adopted in various application areas such as network management, information retrieval, electronic commerce, and planning tasks. As a result, several architectures have been designed for multiagent systems. These architectures help to solve conflicts or other problems between agents.

Among these architectures, we can list the blackboard architecture, the subsumption architecture, the architecture of competitive tasks, the architecture of production systems, the architecture scalable agent-based information retrieval engine (SAIRE), and the agent architecture based on behaviors.

4.7.1 Blackboard Architecture

The blackboard architecture is a kind of "meta-architecture" that is used to implement other architectures (Ferber 1997). It is most commonly used in cognitive multiagent systems. It is developed as part of DAI for the speech recognition system with HEARSAY II (Erman et al. 1980). It is considered a powerful and flexible architecture to implement such mechanisms of reasoning and computation occurring within the agents in the system "daily vehicle—miles of travel" (DVMT) (Lesser and Corkill 1983).

This architecture, as shown in Figure 4.1, consists of three subsystems: the knowledge sources, the shared database, and the control device (Ferber 1997).

1. The knowledge sources (KS). These knowledge sources are independent modules that do not communicate directly, but rather interact indirectly by sharing information. They work on a space with elements that can solve the problems of conflict. These modules are triggered when the configurations of

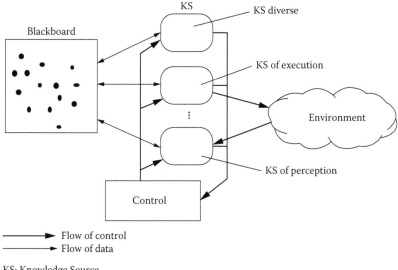

Figure 4.1   System architecture based on blackboard. (Source: Ferber, J. 1997. *Les systèmes multiagents: Vers une intelligence collective.* Paris: InterÉditions.)

blackboard interest them. This occurs due to changes caused by other sources of knowledge.

2. The shared database (or table). This database includes partial states of a problem being resolved, the hypothesis and intermediate results, and information exchanged by the knowledge sources. These bases are decomposed into hierarchies, such as conceptual. This is done in order to structure modeling application domain such as the hypothesis space or solutions.

3. The control device. This device manages access conflicts between sources of knowledge. The problem of control in a blackboard returns to determine what knowledge source is triggered.

This architecture is advantageous, given its flexibility to describe the modules and articulate their operation (Ferber 1997). It is of interest since it is centralized and links between modules are mutable. On the other hand, this architecture is disadvantageous since its inefficiency is due to the high expressiveness of its control. Therefore, this architecture is primarily used for the prototyping phase of a system or when response times are not too restrictive.

4.7.2 Subsumption Architecture

The subsumption architecture is adopted for the constitution of reactive agents (Brooks and Connell 1986). This architecture decomposes an agent into vertical modules where each is responsible for one type of very limited behavior (Ferber 1997). These modules perform their tasks in parallel. The interactions between modules are fixed. They run through a dominance relation defined in the design. However, if two modules are in conflict, then only the information provided by the dominant module is considered. However, if the dominant module is not working and the lower module produces a result, then this will be retained. Figure 4.2 shows the process of dominance by the upper modules that may prevent the release of lower modules.

This architecture is applied in complex systems such as a robot system explorer. It is used to describe reactive agents, as well as cognitive agents, considering that the superior modules are the most reflexive and the lower modules are the most cognitive.

4.7.3 Architecture of Competitive Tasks

The architecture of competitive tasks is composed of agents, each having a set of tasks in which each task will be active at a time (Ferber 1997). These tasks are competing to be elected by a decision mechanism. This mechanism takes into account various parameters such as the weight of the task, the application context, and the information from outside.

This architecture is shown in Figure 4.3. In this architecture, "go find object," "recover energy," or "explore" are the tasks. As well, "move arm" or "move up" are primitive actions. In other words, these actions are considered the words of the language that define the tasks.

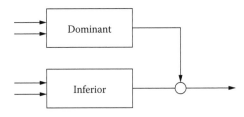

Figure 4.2   Subsumption architecture. (Source: Ferber, J. 1997. *Les systèmes multiagents: Vers une intelligence collective.* Paris: InterÉditions.)

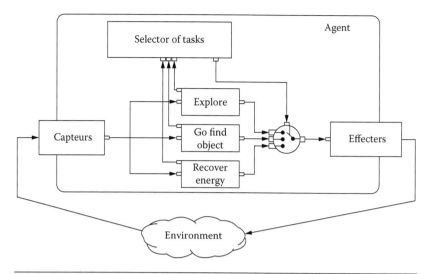

Figure 4.3 System architecture based on competitive tasks. (Source: Ferber, J. 1997. *Les systèmes multiagents: Vers une intelligence collective.* Paris: InterÉditions.)

4.7.4 Architecture of Production Systems

The architecture of production systems is well known in the field of artificial intelligence. As part of a multiagent system, each agent represents a production system as shown in Figure 4.4 (Ferber 1997). The architecture of this system is represented by a combination of a fact base (FB), a base of production rules (BR), and the inference engine (IE) represented as an interpreter, which is provided as the functions of interpretation and execution. The function of perception helps place the perceived functions or the messages within the fact base during the operation of the inference engine. This is intended to allow the base of rules to take them into account directly.

The production rules are defined by a variety of syntaxes, which are usually represented in the form of "If <list-conditions> then <list-actions>." The <list-conditions> is associated with elements of the base of facts and the <list-actions> includes elementary actions such as add or remove items from the base facts.

Some actions can directly activate commands to run the agent. If the conditions of the list-condition of a rule are validated, the list-action of this rule will be executed. However, if several rules are activated simultaneously, there will be conflict. In this case, the control system of the inference engine triggers the rule with the highest

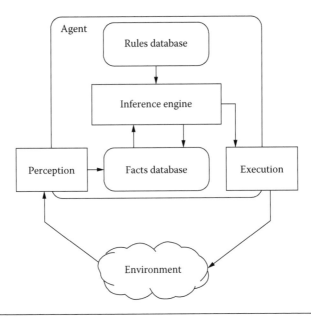

Figure 4.4 Architecture of agent based on production system. (Source: Ferber, J. 1997. *Les systèmes multiagents: Vers une intelligence collective.* Paris: InterÉditions.)

priority. This priority is determined by the internal parameters of the different rules, or the first rule will be executed if several have the same priority level.

This architecture is used in the writing of large knowledge bases. It has two drawbacks (Ferber 1997):

1. The need to know the order of application of rules to be sure that the result is consistent with the objective of the system
2. The noncombination rules due to the differences of procedures and programming functions

To overcome these problems, it is necessary to build packages of rules containing a set of rules related to each other functionally. Therefore, the application of this architecture is reduced and the programs implemented in this way should be tested to check system consistency.

4.7.5 Architecture of Scalable Agent–Based Information Retrieval Engine (SAIRE)

The SAIRE architecture is one of seven projects of digital library technology (DLT) (Odubiyi et al. 1997). This architecture has been established under the program of Information Infrastructure Technology and Applications (IITA) in the National Aeronautics and Space Administration (NASA). It is used for intelligent software agents and the understanding of natural language. As well, it leads to research tools and access to public land data and data relating to the science of space through the Internet.

The architecture of an agent SAIRE is a multiagent architecture in a distributed production system (Figure 4.5) (Das and Kocur 1997). In this architecture, each process C-language integrated production system (CLIPS) is an agent "agent manager" (AM) and one or more specialist agents. The agent (AM) and each specialist agent are implemented as modules. The implementation of modules gives agents their own knowledge bases "knowledge base." Agents in the environment CLIPS are organized in a hierarchical organization. The agent AM is at a level above specialized agents in the hierarchy. The hierarchical organization is recommended since it allows each agent AM to manage any specialist agent in its environment CLIPS.

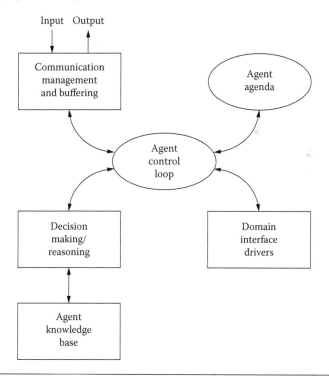

Figure 4.5 Architecture of SAIRE agent. (Source: Das, B., and Kocur, D. 1997. In *IEEE Knowledge and Data Engineering Exchange Workshop (KDEX '97)*, 27–35.)

4.7.6 Architecture of Agents Based on Behavior

The architecture of agents based on behavior is a complementary combination between planning systems and systems of deliberation (Chaib-Draa, Paquet, and Lamontagne 1993). This architecture is inspired by the cognitive work based on behavior (Rasmussen 1986).

This architecture is reactive. It responds quickly to sudden changes. It also enables the planning of acts achieving a goal or task. The latter are either identified or recognized or directly perceived by the agent when in familiar situations. However, in the case of unfamiliar situations, this architecture allows decision making for each participant.

This architecture is described, as shown in Figure 4.6, as follows:

- The agent perceives one or more pieces of information from its environment.
- This information is pushed to act if it is directly perceived as an action, or plan if it is perceived as a task or goal.
- On the other side, if the information is perceived as neither of these forms, the agent should identify or recognize a situation. The identification and recognition permit the agent

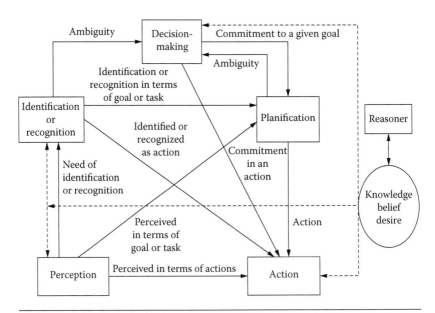

Figure 4.6 Architecture of agent based on behaviors. (Source: Chaib-Draa, B., and Vanderveken, D. 1999. In *Intelligent Agents V. Agent Theories, Architectures and Language*, ed. Müller, J. P., Singh, M. P., and Rao, A. S., 1555:363–379. Berlin: Verlag.)

to act if the situation is identified or recognized in terms of action or to plan if the situation is identified or recognized in terms of goal or task.

- If the agent is faced with an ambiguity or if it is necessary to choose a goal against an unusual situation, it should make decisions leading to a given goal.

This architecture takes into account the intentions of other agents. Social knowledge such as social standards and cooperative rules is integrated. This aims to enrich the levels of routine and familiar situations by providing certain coordination with other agents.

4.8 Conclusion

In this chapter we detailed the different types of agents and their properties. Subsequently, we presented the communication between agents and acts of language and of conversation proposed to define languages and protocols between the different agents of the group. Then, we described the cooperation (i.e., interaction, coordination, and negotiation) of agents and planning in a multiagent system to solve problems of conflict. Finally, we examined the architecture of agents to choose an efficient architecture for our model.

We can consider that the integration of agents in a distributed multiagent system allows, in the presence of complex problems, managing the actions in a well-organized way.

Conclusion of Part 1

In summary, we focused our study on the biometric iris of the eye, which is considered a reliable defense and powerful for securing access to information in the network.

We presented the different methods used to improve the algorithm of iris recognition developed by John Daugman. These methods are numerous, especially at the level of the localization of external and internal edges of the iris, and at the level of eliminating the effects of upper and lower eyelids. It remains to be seen which method gives the best performance in this algorithm. This is the subject of the next part.

The use of existing cryptographic techniques is a response to the security of information transmitted over the network. The integration of multiagent systems has shown effectiveness for the treatment of complex problems in various areas of systems users.

PART 2

CRITICAL ANALYSIS OF METHODS OF IRIS RECOGNITION

The diversity of methods used to improve the algorithm of iris recognition has led us to detail them to assess the effectiveness of each. We are interested in methods for the localization of external and internal edges of the iris, as well as the methods of eliminating the effects of the upper and lower eyelids.

5

EXISTING METHODS FOR LOCALIZATION OF EXTERNAL AND INTERNAL EDGES OF THE IRIS OF THE EYE

This chapter aims to present the different existing methods used for the localization of external and internal edges of the iris of the left or right eye. We will detail five simulations of these methods. A comparative analysis and an evaluation of the results of these simulations are presented in the Section 5.3. After analyzing these results, we will propose in Chapter 7 of Part 3 a new model that is more efficient at the level of processing time.

5.1 Introduction

As we detailed in Chapter 2, John Daugman was the first to propose the method of recognition based on the iris of the eye. He used the integrodifferential operator to locate the external and internal edges of the iris (Daugman 2004). This method has also been used by Huang et al. (2002).

Several researchers have worked to improve the algorithm developed by John Daugman at the level of the localization phase of external and internal edges of the iris (cf. Chapter 2, Section 2.5.2); Daouk et al. (2002) adopted the circular Hough transform to improve the phase of localization of external and internal edges of the iris. Tian et al. (2006) used the method of research of the square (cf. Chapter 2, Section 2.5.2) to detect the region of the pupil. However, Miyazawa et al. (2005) have located the external and internal edges by elliptical Hough.

5.2 Tests and Simulations

In this section we present five simulations implementing methods with the objective of localization of external and internal edges of the iris. Our tests were conducted on a sample of 257 images of three types:

1. Random images from Internet documents
2. Images taken by the Sony CCD camera with a resolution of 520 TV lines
3. Images from the CASIA iris image database (CASIA 2006)

These different types of images are illustrated, respectively, by Figures 5.1–5.5(a), (b), and (c).

We used MATLAB® R2006a on a Pentium IV with a 2.2 GHz processor and 1 MB of memory to perform these simulations.

5.2.1 First Simulation: Method of Daugman and Huang

This simulation is based on the methodology adopted by Daugman (2004) and Huang et al. (2002). This method consists of defining a circle of radius r and center (x_0, y_0) corresponding to the first maximum value of the summation of the intensities of the points that constitute it. This approach is based on the principle of the integrodifferential operator. It is applied to the localization of the external and internal edges of the iris.

The average time for this process is 18 seconds. The result of this simulation is presented in Figure 5.1.

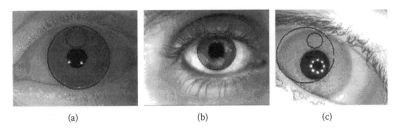

(a)   (b)   (c)

Figure 5.1 Localization of external and internal boundaries by the sum of the maximum intensities.

5.2.2 Second Simulation: Method of Daugman and Huang and Tian

This simulation is based on the same method as that adopted in the first simulation for the localization of the outer boundary of the iris. For the localization of the internal edge of the iris, the test turns on the determination of the circle defining the region with a minimum average of gray (i.e., the blackest region of the pupil) (Tian et al. 2006).

The average time for this process is 16 seconds. The result of this simulation is illustrated in Figure 5.2.

5.2.3 Third Simulation: Method of Daouk and Tian

This simulation is based on the application of the circular Hough transform used by Daouk et al. (2002) and Tian et al. (2004) for the localization of the external edge of the iris. The localization of the internal edge of the iris uses the method of the second simulation.

The average time for this process is 14.5 seconds. The result of this simulation is presented in Figure 5.3.

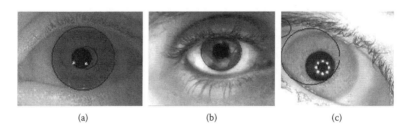

(a) (b) (c)

Figure 5.2 Localization of external boundary by the sum of the maximum intensity, and of the internal edge by limiting the blackest region in the image.

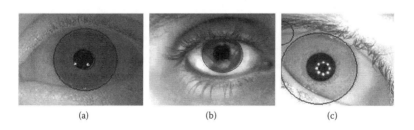

(a) (b) (c)

Figure 5.3 Localization of external edge by circular Hough transform, and internal edge by limiting the darkest region in the image.

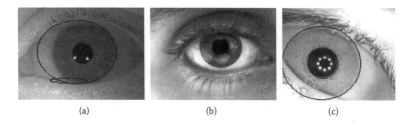

Figure 5.4   Localization of external and internal edges by the elliptical Hough transform.

5.2.4 Fourth Simulation: Method of Miyazawa

This simulation is based on the application of the elliptical Hough for the localization of external and internal edges of the iris (Miyazawa et al. 2005). This transform is considered as a generalized circular Hough transform.

The average time for this process is 200 seconds. The result of this simulation is illustrated in Figure 5.4.

5.2.5 Fifth Simulation: Method of Daouk and Tian

This simulation is based on the application of the circular Hough transform, used in the third simulation, for the localization of the external and internal edges of the iris (Daouk et al. 2002; Tian et al. 2004).

The average time for this process is 13.5 seconds. The result of this simulation is presented in Figure 5.5.

5.3 Analysis of Simulations

Our study focuses on the iris of the eye, which is represented geometrically by a ring bounded by two conical (i.e., circle or ellipse) edges C1 and C2, as shown in Figure 5.6, including:

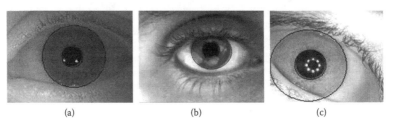

Figure 5.5   Localization of the external and the internal edges by the circular Hough transform.

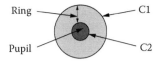

Figure 5.6 Geometric representation of the ring.

- C1: external edge of the ring
- C2: internal edge of the ring

In the case of an ellipse, the two edges C1 and C2 constitute the perimeter of the ellipse defined by the triplet (a, b, c), including:

- a: small radius (a ∈ ℝ)
- b: large radius (b ∈ ℝ)
- c: center defined by its coordinates (x0, y0) in a Cartesian system

In the case of a circle, the parameters are just the data of a couple (r, c), including:

- r: radius (r ∈ ℝ)
- c: center defined by its coordinates (x, y) in a Cartesian system

In the case of an ellipse defined by (a, b, c), where a = b = r, this conical edge is not other than a circle.

For a given image, we focus on internal or external edges of the iris and their localization (cf. Chapter 2, Section 2.6). We recall that the localization of a conic is characterized by the parameters (a, b) and the center of the conic edge in the image. The Cartesian equation is

$$\frac{(x-x_0)^2}{a^2} + \frac{(y-y_0)^2}{b^2} = 1$$

Figure 5.7 illustrates these parameters.

Consider a finite image F composed of a product of three different families (F_1, F_2, F_3).

$$F = F_1 \times F_2 \times F_3$$

- F1 is a set of 35 images taken at random from the Internet documents.

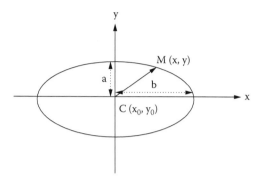

Figure 5.7   Geometric representation of the ellipse.

- F2 is a set of 72 images taken by the Sony CCD color camera with a resolution of 520 TV lines.
- F3 is a set of 150 images issued from the CASIA iris image database (Miyazawa et al. 2005).

For each image belonging to the families Fj, we will determine the external edge C1 and the internal edge C2 previously defined.

Both edges are considered as a circle in the methods used in simulation 1, simulation 2, simulation 3, and simulation 5. This circle will be characterized by the parameters $(r_i, c_i(x_i,y_i))$ for C1 and $(r_p, c_p(x_p,y_p))$ for C2.

Both edges are considered as ellipses in the method used in simulation 4. This ellipse is characterized by the parameters $(a_i, b_i, c_i(x_i,y_i))$ for C1 and $(a_p, b_p, c_p(x_p,y_p))$ for C2.

Let

- n_1 represent the cardinality of F_1, n_1 = card (F_1)
- n_2 represent the cardinality of F_2, n_2 = card (F_2)
- n_3 represent the cardinality of F_3, n_3 = card (F_3)

5.3.1 Concentricity of Edges C1 and C2

The importance of the concentricity of edges C1 and C2 led us to evaluate the distance between these two centers $c_i(x_i,y_i)$ and $c_p(x_p,y_p)$. This distance is normalized on a scale from 0 to 100 to evaluate one method versus another.

We set $d_{j,k}$ as the distance between these two centers $c_i(x_i, y_i)$ and $c_p(x_p, y_p)$ for a method M_k ($k = 1, \ldots, 5$) and belonging to the family Fj ($j = 1, \ldots, 3$).

This distance is given by

$$d_{j,k} = |x_i - x_p| + |y_i - y_p|$$

We take $d_{j,k}^{\min}$ as the minimum value of $d_{j,k}$.

We consider $D_j^{\max}$ the maximum value of $d_{j,k}^{\min}$ for a family Fj.

$$D_j^{\max} = \underset{k=1,\ldots 5}{\text{MAX}}\, d_{j,k}^{\min}$$

We bring the value of $d_{j,k}^{\min}$ to the scale from 0 to 100 rounded to the nearest integer $D_{j,k}$:

$$D_{j,k} = 100 - \left(\frac{d_{j,k}^{\min}}{D_j^{\max}} \times 100 \right)$$

We obtain Table 5.1 with the values of $D_{j,k}$ based on our sample of families F_1, F_2, and F_3 with

- Card $(F_1) = 35$
- Card $(F_2) = 72$
- Card $(F_3) = 150$

The corresponding histogram is seen in Figure 5.8.

We can see that the higher the value of $D_{j,k}$ is, the fewer the pixels will be rejected in the analysis.

Given the obtained results, we note that the value $D_{j,k}$ of the method M_5 does not have a big difference compared to the methods M_2 and M_3 of the family F_1. On the other side, the method M_1 is clearly wrong compared to other methods of this family since its value $D_{j,k}$ is equal to zero.

Table 5.1   Table of Values $D_{j,k}$

	F_1	F_2	F_3
M_1	0	71	0
M_2	91	95	77
M_3	91	87	86
M_4	85	0	91
M_5	94	95	100

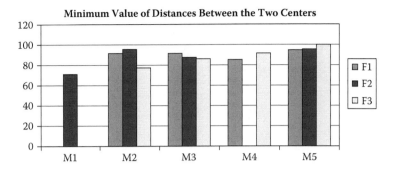

Figure 5.8   Rate of values $D_{j,k}$ relative to each method on the three families.

For the family F_2, both methods M_5 and M_2 have the same value $D_{j,k}$. They are considered the best compared to other methods of this family. On the other hand, the method M_4 is clearly wrong compared to other methods of this family since its value $D_{j,k}$ is equal to zero.

At the level of the family F_3, the method M_5 is the best compared to other methods since its value $D_{j,k}$ is equal to 100. On the other side, M_1 is clearly wrong with a value of $D_{j,k}$ equal to zero.

Therefore, we can conclude that the method M_5 is the best of all families in terms of concentricity between the two edges C1 and C2.

5.3.2 Temporal Aspect

The time efficiency is to compare the effectiveness of a method M_k ($k = 1, \ldots, 5$) relative to another of the three families F_1, F_2, and F_3. This efficiency will be calculated later.

We set $t_{h,k}^j$ the computation time to determine the edges C1 and C2 of the iris number h belonging to the family Fj ($j = 1, \ldots, 3$) and the method M_k.

This gives us an estimation of average time:

$$\overline{T}_{j,k} = \frac{1}{n_j} \sum_{h=1}^{n_j} t_{h,k}^j$$

We set

$$\overline{T}_j = \sum_{k=1}^{5} \overline{T}_{j,k}$$

The calculation of the time efficiency of the method M_k on the family Fj for determining the edges C1 and C2 is

$$T_{j,k}^e = 1 - \frac{\overline{T}_{j,k}}{\overline{T}_j}$$

We get Table 5.2 of time efficiency based on our sample of the families F_1, F_2, and F_3, with

- Card $(F_1) = 35$
- Card $(F_2) = 72$
- Card $(F_3) = 150$

The corresponding histogram is seen in Figure 5.9.

We can see that the higher the rate of time efficiency, the more efficient the method is in terms of computation time for the localization of the edges C1 and C2.

Given the obtained results at the level of time efficiency, the method M_5 does not have a big difference compared to methods M_1, M_2, and M_3 ($T_{j,k}^e = 0.97$) for the localization of the edges C1 and C2 in the

Table 5.2   Table of Values $T_{j,k}^e$

	F_1	F_2	F_3
M_1	0.96	0.93	0.92
M_2	0.96	0.93	0.93
M_3	0.95	0.94	0.94
M_4	0.15	0.26	0.26
M_5	0.97	0.94	0.94

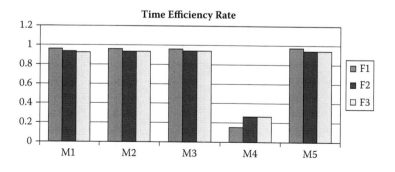

Figure 5.9   Time efficiency rate relative to each method on the three families.

three families F_1, F_2, and F_3. On the other side, the method M_4 is clearly less effective compared to other methods in all families.

5.3.3 Length of Radius of Edge C1

Our objective is to characterize the method that gives the best length of the radius of the edge C1 to be standardized in a scale from 0 to 100 to evaluate one method versus another.

We take $R_{j,k}^{max}$ to be the maximum value of the radius r of the edge C1 by family Fj ($j = 1, \ldots, 3$) for each method M_k ($k = 1, \ldots, 5$).

Note that in the case of an ellipse with radii a and b, we choose as value r of the radius of the edge C1 the maximum value of a and b:

$$r = Max\,(a,b)$$

We calculate R_j^{max} as the maximum of the radius of the edge C1 by family Fj.

We bring the value of $R_{j,k}^{max}$ to the scale from 0 to 100 rounded to the nearest integer $R_{j,k}$:

$$R_{j,k} = \frac{R_{j,k}^{max}}{R_j^{max}} \times 100$$

We obtain Table 5.3 of values $R_{j,k}$ based on our samples in the families F_1, F_2, and F_3, with

- Card (F_1) = 35
- Card (F_2) = 72
- Card (F_3) = 150

The corresponding histogram is seen in Figure 5.10.

We can see that the more important the value of $R_{j,k}$ is, the more pixels we have to be processed.

Table 5.3   Table of Values $R_{j,k}$

	F_1	F_2	F_3
M_1	100	84	89
M_2	100	84	89
M_3	100	100	98
M_4	100	95	100
M_5	100	100	98

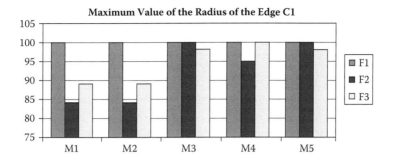

Figure 5.10   Rate of values $R_{j,k}$ relative to each method on the three families.

Given the obtained results, the five methods—M_1, M_2, M_3, M_4, and M_5—have the same value (equal to 100) for the family F_1.

For the family F_2, the best value $R_{j,k}$ (equal to 100) is obtained for both methods M_3 and M_5. Consequently, these two methods are considered the best compared to other methods in this family.

At the level of the family F_3, the method M_4 is better than the two methods M_3 and M_5, but it is clearly much better compared to the other two methods, M_1 and M_2.

5.3.4 Comparison of the Methods by Outranking Relation

We will compare the effectiveness of the five methods using outranking relations. Therefore, we implement the method Electre I of Bernard (1985) for each set F_j ($j = 1, \ldots, 3$). This method is based on the concept of multicriteria outranking.

The criteria of comparison, detailed before, focus on

- The minimum distance between the centers of the two edges C1 and C2 relative to the maximum value of $D_{j,k}$
- The rate of time efficiency $T^e_{j,k}$
- The maximum length of the radii of the edges C1 relative to the maximum value of $R_{j,k}$

The values of the previously listed criteria will be marked on a scale from 0 to 100.

Based on our three samples F_1, F_2, and F_3 we obtain Tables 5.4–5.6.

Table 5.4   Table of Values $D_{j,k}$, $T^e_{j,k}$, and $R_{j,k}$ for the Family F_1

F_1	$D_{J,K}$	$T^e_{j,k}$	$R_{J,K}$
M_1	0	0.96	100
M_2	91	0.96	100
M_3	91	0.95	100
M_4	85	0.15	100
M_5	94	0.97	100

Table 5.5   Table of Values $D_{j,k}$, $T^e_{j,k}$, and $R_{j,k}$ for the Family F_2

F_2	$D_{J,K}$	$T^e_{j,k}$	$R_{J,K}$
M_1	71	0.93	84
M_2	95	0.93	84
M_3	87	0.94	100
M_4	0	0.26	95
M_5	95	0.94	100

Table 5.6   Table of Values $D_{j,k}$, $T^e_{j,k}$, and $R_{j,k}$ for the Family F_3

F_3	$D_{J,K}$	$T^e_{j,k}$	$R_{J,K}$
M_1	0	0.92	89
M_2	77	0.93	89
M_3	86	0.94	98
M_4	91	0.26	100
M_5	100	0.94	98

We will choose a weight on each of the criteria $D_{j,k}$, $T^e_{j,k}$, and $R_{j,k}$. We proposed to take the following as weights:

- $P_1 = 0.70$ for the criterion $D_{j,k}$
- $P_2 = 0.25$ for the criterion $T^e_{j,k}$
- $P_3 = 0.05$ for the criterion $R_{j,k}$

We explain the choice of the weights; we assess that the criterion $D_{j,k}$ is more important than the other criteria for determining a better localization of the edges C1 and C2. The criterion $T^e_{j,k}$ comes in second place because it is important to gain more time to locate the edges C1 and C2. The criterion $R_{j,k}$ is of little importance compared to other criteria, since the radius of the edge C1 varies from one to another iris. In addition, this criterion can have a maximum value, but for a bad localization of the edge C1.

We obtain an outranking relation that is set according to the thresholds of concordance p and discordance q as the method of Electre (Bernard 1985).

Recall that from the notion of outranking of an action $M_{k'}(k' = 1, \ldots, 5)$ by an action $M_k(k = 1, \ldots, 5; k \neq k')$ emerge two concepts (Bernard 1985):

1. The concept of concordance expresses that a majority of criteria are in favor of a method M_k.
2. The concept of discordance reflects the fact that there can be, for the criteria which do not belong to this majority, a too sharp pressure in favor of a method $M_{k'}$.

We conducted the following simulations:

$$p = 0.75$$

and

$$q = 0.2$$

This gives for each of the three families the graphs G_1, G_2, and G_3 of outranking as seen in Figures 5.11–5.13.

In the simulation of the family F_1, as shown in Figure 5.11, the method M_5 outclasses the methods M_2, M_3, M_4, and M_1. Both methods M_2 are M_3 are identified in a method M' in the graph

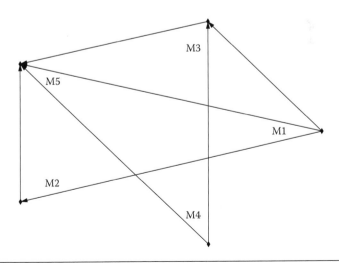

Figure 5.11   Graph G1 of outranking of the family F_1.

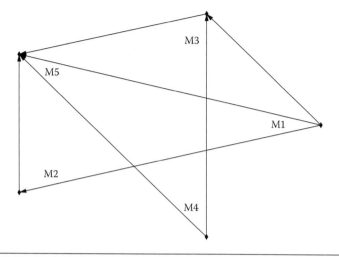

Figure 5.12   Graph G2 of outranking of the family F_2.

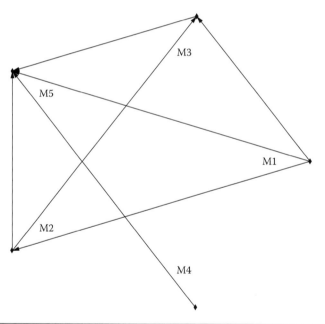

Figure 5.13   Graph G3 of outranking of the family F_3.

because they are connected by a circuit of the graph. The method M′ outclasses the methods M_4 and M_1, while the methods M_4 and M_1 do not outclass any method. In consequence, the method M_5 is not dominated by any method and is a core element (Bernard 1985) of the graph of outranking G_1.

In the simulation of the family F_2, as shown in Figure 5.12, the methods M_5 and M_2 outclass the method M_1. The method M_5 outclasses the methods M_2, M_3, and M_4. The method M_3 outclasses the methods M_4 and M_1. The methods M_4 and M_1 do not outclass any method. In consequence, the method M_5 is not dominated by any method and is a core element of the graph of outranking G_2.

In the simulation of the family F_3, as shown in Figure 5.13, the method M_5 outclasses the methods M_2, M_3, M_4, and M_1. The method M_2 outclasses the method M_1. The method M_3 outclasses the methods M_2 and M_1. The methods M_4 and M_1 do not outclass any method. In consequence, the method M_5 is not dominated by any method and is the core element of the graph of outranking G3.

We find, based on the concept of the method Electre I, that the method M_5 has the best localization, since in the three previous simulations this method dominated all other methods and was not dominated by any. We can deduce that the application of the circular Hough transform is most suited to the model that we will develop later in this book.

5.4 Conclusion

In this chapter we have detailed the different methods presently used for the localization of the external and internal edges of the iris of the eye. These methods have been adopted by several researchers. We have presented the simulations on sample images. At the end of these simulations, we will propose in Chapter 7 of Part 3 a new, appropriate, and effective model in terms of processing time for the localization of external and internal edges of the iris.

In the case of a model of recognition based on the iris of the eye, the upper and lower eyelids can hide a part of the iris. This causes a problem for the accuracy at the level of authentication. We will present in the next chapter the different methods used to solve this problem.

6

EXISTING METHODS FOR ELIMINATING EFFECTS OF THE EYELIDS

In this chapter we propose to present the existing methods used for the elimination of the effects of upper and lower eyelids that can hide a part of the iris of the eye. We will detail three simulations of these methods. Based on these approaches, in Chapter 7 we will present our most effective method for such treatments.

6.1 Introduction

Normalization is an important phase in the process of authentication. In addition to the functions detailed in previous chapters (cf. Chapter 2, Section 2.5.3), this phase leads to eliminating the effects of upper and lower eyelids. Several researchers have proposed various methods to solve these problems (cf. Chapter 2, Section 2.5.6). We present three simulations based on different concepts currently used.

6.2 Tests and Simulations

Recall that our tests are performed on a sample of 257 images taken from the CASIA iris image database (CASIA 2006). We used MATLAB R2006a on a Pentium IV with a 2.2 GHz processor and 1 MB of memory to perform these simulations.

The three following simulations were performed on four samples of images:

1. The left iris of person 1
2. The left iris of person 1 with a position different from the preceding

3. The right iris of person 1
4. The right iris of person 2

The binary gabarits (cf. Chapter 2, Section 2.5.4) issued from these images will be represented in the three following simulations by gabarit(a), gabarit(b), gabarit(c), and gabarit(d).

The results of verification between two binary gabarits are obtained using the Hamming distance (HD) (Daouk et al. 2002).

These results show the dissimilarity rate (i.e., the density of the dissimilar points) between two given binary gabarits. Recall that we also used the discrete Haar transform in two dimensions for the feature extraction of the iris gabarit (cf. Chapter 2, Section 2.5.4).

6.2.1 First Simulation: Method of Daugman and Wildes

This simulation is based on the concept of Daugman (2004) and Wildes (1997). They modeled the upper and lower eyelids with parabolic arcs as shown in Figure 6.1(a). These parabolic arcs are defined by parameter "splines" determined by methods of statistical estimation (Daugman 2004).

Recall that a spline is a piecewise function by polynomials (Demengel and Pouget 1998). For interpolation, spline is preferred in an interpolation polynomial. Splines are used to approximate complex edges. Note that the determination of a spline in a digital image is based on the size of the form to define, the luminance, and the color of the pixel.

The gabarit shown in Figure 6.1(b) is composed of the region of the iris isolated from the given image. The results of various combinations

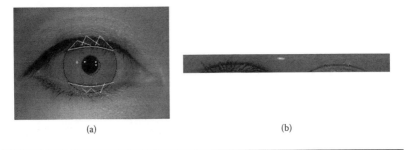

(a)   (b)

Figure 6.1   Result of simulation 1 by eliminating the region of the iris hidden by the eyelids in the form of parabolic arcs.

of binary gabarits, extracted from the four sample images, are the following:

- Rate of dissimilarity (gabarit(a), gabarit(b)) = 0.0239
- Rate of dissimilarity (gabarit(a), gabarit(c)) = 0.0241
- Rate of dissimilarity (gabarit(a), gabarit(d)) = 0.0243
- Rate of dissimilarity (gabarit(c), gabarit(d)) = 0.0245

6.2.2 Second Simulation: Method of Myazawa and Daouk

In this simulation, we used the method proposed by Myazawa et al. (2005) and Daouk et al. (2002). This method consists of taking the lower part of the iris for the authentication, as shown in Figure 6.2(a).

The gabarit is represented in Figure 6.2(b). The results of various combinations of binary gabarits, extracted from the four sample images, are the following:

- Rate of dissimilarity (gabarit(a), gabarit(b)) = 0.0183
- Rate of dissimilarity (gabarit(a), gabarit(c)) = 0.0159
- Rate of dissimilarity (gabarit(a), gabarit(d)) = 0.0183
- Rate of dissimilarity (gabarit(c), gabarit(d)) = 0.0185

6.2.3 Third Simulation: Method of Tian

This simulation is based on the concept of Tian et al. (2006). They adopted the linear Hough transform on the image. They used the model with three lines [1p] to close the eyelid for each eyelid's edge in order to form the gabarit shown in Figure 6.3(b).

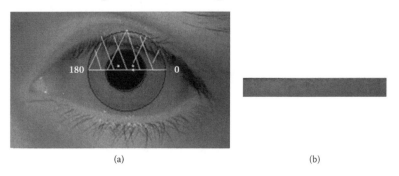

(a)   (b)

Figure 6.2   Result of simulation 2 by eliminating half the upper part of the iris.

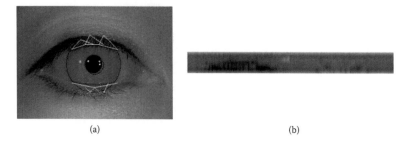

(a)   (b)

Figure 6.3   Result of simulation 3 by eliminating the region of the iris hidden by the eyelids in the form of three straight segments.

We find that this method is to determine the one or two or three line segments cutting the external edge of the iris and delimiting the region of the eyelid as shown in Figure 6.3(a).

The results of the rate of the dissimilarity of various combinations of binary gabarits, extracted from four samples of images, are as follows:

- Rate of dissimilarity (gabarit(a), gabarit(b)) = 0.0180
- Rate of dissimilarity (gabarit(a), gabarit(c)) = 0.0154
- Rate of dissimilarity (gabarit(a), gabarit(d)) = 0.0180
- Rate of dissimilarity (gabarit(c), gabarit(d)) = 0.0176

6.3 Evaluations of Simulations

The simulations we conducted show that the thresholds from which images are differentiated have a very low value. Indeed, in the first simulation we find that the threshold is 0.023. However, Daugman recommends taking a threshold value of 0.32. Therefore, with the threshold of Daugman, all images having a threshold between 0.32 and 0.023 are considered identical.

In the second simulation, we find that the results obtained by the method of Miyazawa et al. (2005) and Daouk et al. (2002) give the same rate of frequency of dissimilarity (equal to 0.0183) in two identical or different irises. Both rates of dissimilarity (equal to 0.0159) in two different irises are smaller than that for the same two irises (equal to 0.0183). The binary method (cf. Chapter 2, Section 2.5.4) applied to the lower part of the iris leads to very poor results.

At the level of the third simulation, we have similar observations as those of simulation 2. The frequency value of dissimilarity on two identical or different irises is equal to 0.0180. It is very close to the value obtained in simulation 2. As well, the rate value of dissimilarity (0.0154 or 0.0176) on two different irises is less than the one on two identical irises (0.0180).

Following these evaluations, we propose in Chapter 7 to improve the method of elimination of the effects of the eyelids and refine a better estimation of the threshold.

6.4 Conclusion

In this chapter we have detailed the different existing methods used to eliminate the effects of the upper and lower eyelids based on three simulations. An evaluation of each of these methods has shown that their results are not favorable. This justifies our proposed method in Chapter 7 of the third party, as well as the estimated threshold value.

Conclusion of Part 2

In conclusion, the method based on the circular Hough transform has brought effectiveness at processing time and accuracy for the localization of external and internal edges of the iris. This method is most suitable for our model that we propose in Chapter 7 of Part 3. On the other hand, the methods used to eliminate the effects of upper and lower eyelids gave negative results.

Our discussion will turn to conceiving an effective method for eliminating the effects of the upper and lower eyelids. This method will be detailed in Chapter 7 and the justification of our choice will be discussed in Chapter 10.

PART 3

OUR PROPOSED MODEL: THE IRIS CRYPTO AGENT SYSTEM

This part aims to present our model, called the IrisCryptoAgentSystem (ICAS). This model is divided into two parts: the model of biometric authentication based on the iris of the eye and transmission of encrypted data over the Internet, and the model multiagent system (MAS). ICAS should secure access to information and ensure the protection of confidential information. Our thinking carries on the possibilities to implement new methods in the biometric authentication model that will increase the efficiency of our ICAS. Our approach is based on the improvement of existing methods.

7

BIOMETRIC MODEL FOR AUTHENTICATION USING THE IRIS OF THE EYE

The proposed model described in this chapter is based on the biometric authentication method using the iris of the eye and the asymmetric cryptography using the Rivest–Shamir–Adleman (RSA) algorithm (Stallings 1999). This model is founded on the use of biometric iris signatures' gabarit of a person to access a computer system. The justification for the choice of the biometric model based on the iris of the eye is presented in this chapter, as well as the description of the different modules constituting this model. In this model, the authentication process allows one to obtain a diagonal horizontal vertical approximation (DHVA)* encrypted gabarit. This will be compared with other DHVA encrypted gabarits stored in the database to verify access to the computer system. A method of classification of these gabarits is detailed in order to speed the verification process, especially for a large database.

7.1 Introduction

The objective of the proposed model is to secure access to confidential information. This model is based on the biometric signature of the iris of the eye gabarit (cf. Chapter 2, Section 2.5) and the method of asymmetric cryptography using the RSA algorithm (Stallings 1999). This choice is justified in the next section. This model defines

1. The process of determining the gabarit DHVA
2. The coding of the gabarit DHVA using the RSA algorithm

---

* This name is inspired by the concept of approximation of the Haar wavelet.

3. The classification of the gabarit DHVA to be recorded in specific partitions in the database

The determination of the gabarit DHVA represents the process to obtain the biometric characteristics of the iris of the eye for a given person. The encoding of the gabarit DHVA consists of storing the data, in a secure way, in the database in order to make it difficult to recover. The classification of the gabarit DHVA aims to accelerate the verification process. This phase leads to search with an optimal time for the gabarit DHVA, stored in the database, to be compared with the required gabarit DHVA for the given person.

7.2 Justification of the Choice of the Proposed Biometric Model

The proposed model is based on the biometric method of iris scanning applied to the asymmetric cryptographic method using the RSA algorithm (Stallings 1999). The biometric method of iris recognition is more efficient and accurate than other methods currently used for secure access to data, since the error rate is minimal, of the order of $1/1.2 \times 10^6$ (Rosistem 2001), as shown in Table 7.1. Moreover, iris patterns are formed during the first 2 years of life and remain stable (Perronnin and Dugelay 2002).

According to Bron et al. (1997), irises are unique because of their chaotic morphogenesis (i.e., asymmetry). Both irises for the same person are different, including twins. The iris is not changed by contact lenses or glasses. Similarly, it is not affected by cataracts or age. According to Meng and Xu (2006), the probability that two irises are identical is statistically estimated to be 1 in $2^{173} \approx 8.352 \times 10^{-53}$.

Table 7.1   Probability of Duplication and Level of Security of Different Biometric Techniques

TECHNOLOGY	PROBABILITY OF DUPLICATION	LEVEL OF SECURITY
Voice recognition	$1/3 \times 10$	Low
Facial recognition	$1/10^2$	Low
Signature recognition	$1/10^2$	Low
Fingerprint recognition	$1/10^3$	Average
Recognition of hand shape	$1/7 \times 10^2$	Low
Iris recognition	$1/1.2 \times 10^6$	High

In addition to the efficiency and accuracy, the method of iris recognition is considered the most efficient compared to other biometric methods (cf. Chapter 2, Section 2.5). It is not impossible to imitate a fingerprint or a face. As far as signature, voice and hand geometry suffer from a very low level of security. However, the characteristics of the iris are difficult to forge since they cannot be modified by surgery.

The second method used in our model is asymmetric cryptography using the RSA algorithm, which reinforces the security level of data access (e.g., iris biometric characteristics or data transmitted through the network). This technique is considered to be robust in terms of speed of execution of an operation. As well, it is difficult to know the algorithm of encryption or decryption and to find the private key (Stallings 1999).

7.3 Description of the Biometric Model

Our biometric model leads to the authentication of users accessing a computer system using the iris of the eye. The person will be identified by personal information (e.g., name, father's name, mother's name, surname, date of birth, gender, address, phone, e-mail) and the left or right iris.

The biometric model of iris recognition is composed of five phases (as shown in Figure 7.1):

1. Phase 1: acquisition of the image
2. Phase 2: manipulation of the image
3. Phase 3: image processing
4. Phase 4: encryption of the gabarit
5. Phase 5: verification

Phase 1, acquisition of the image, aims to capture the image of the eye of the person by a camera. The implementation of this camera will be associated with a certain procedure (e.g., the positioning in front of the camera and the distance from the eye to the camera, etc.) to obtain a good quality image.

Phase 2, manipulation of the image, consists of reducing the image at a certain scale to accelerate the execution time of the following processes, as well as the transformation to grayscale to obtain a binary image.

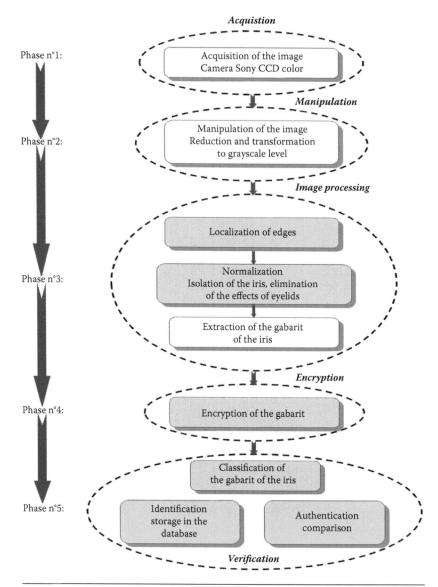

Figure 7.1   Decomposition of the biometric model of the authentication by the iris of the eye.

Phase 3, image processing, is composed of three steps:

1. The localization of the external and internal edges of the iris in the required image
2. The normalization that focuses on the isolation of the iris image and elimination of the effects of the eyelids that can hide a part of the iris

3. The extraction of the gabarit, which consists in extracting biometric features (or biometric signature) of the iris representing the DHVA

Phase 4, encryption of the gabarit, serves to encrypt the gabarit DHVA to be stored in a database in a secure way; otherwise, it would be possible to recover easily.

Phase 5, verification, has two modes:

1. The identification mode of a person that allows this person to access a computer system: it comes to storing personal information of that person and his or her gabarit DHVA encrypted using a method of classification detail later.
2. The authentication mode of a person wishing to access a computer system, which leads to checking right of access: this is to compare the required gabarit DHVA with the gabarit DHVA stored in the database and uses a research method developed later in this chapter.

7.4 Representation of the Different Phases Constituting the Model

Our biometric model is based on a geometric modeling of the iris of the eye, as well as on the elements describing the acquisition of the image and the encryption of the information. As detailed in Section 7.3, our biometric model of iris recognition is composed of five phases. Each phase constitutes an entity with its own specific standards. We will implement the realization of these phases representing the biometric model.

7.4.1 Phase 1—Acquisition of the Image

The images are taken by a Sony CCD color camera with a resolution of 520 TV lines. This type of camera serves to locate the external and internal edges of the iris easily, regardless of its color or illumination scale (i.e., daylight, half darkness, and darkness). To rationalize a proper positioning of the eye in front of the camera and to obtain a usable image, our model focuses on the following elements:

- The image taken, I, is stored in JPEG to gain more detail (Figure 7.2 shows the image quality).
- The distance, d, from the eye to the camera should be between 5 and 10 cm.
- The angle α should be less than 60°.
- The projectors should be installed next to the camera without directly illuminating the eye to keep the ambient illumination intensity stable.

The first three points are considered by most researchers. We have adopted the fourth, concerning the installation of projectors, in order to have better lighting.

We also determined the value of the distance, d, in terms of the taken image, I, because the current system camera is not equipped with a corrective adjustment image such as autofocus. This distance will be established based on the region of the pupil that is the darkest part in the image (the closer we get to the camera the larger is the region of the pupil). This distance d consists in estimating the parameters (ri1, ri2, rp1, and rp2) to locate both external and internal edges of the iris:

- ri1 is the lower bound and ri2 the upper bound of the interval [ri1, ri2] of the search of the radius ρ_{iris} of the external edge C1 of the iris.
- rp1 is the lower bound and rp2 the upper bound of the interval [rp1, rp2] of the search of the radius ρ_{pupil} of the internal edge C2 of the iris.

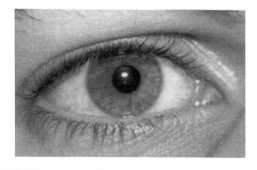

Figure 7.2   Image I captured by the camera Sony CCD color.

7.4.2 Phase 2—Manipulation of the Image

The manipulation of the image consists in reducing the size of the image I to the scale of the quarter I′ of the initial size in order to minimize the detection time of edges. Then, the sample image I′ is transformed from the RGB to the grayscale mode. We choose to do a lossless reduction of the image in order to expedite the edge detection process. As a result, we get a binary image in grayscale level I_g as shown in Figure 7.3.

The grayscale transformation is adopted in all existing models of iris recognition. Reducing the image scale to the quarter is used by Daouk et al. (2002).

7.4.3 Phase 3—Treatment of the Image

The image processing includes the localization of the external and internal edges of the iris, the normalization of the region of the iris, and the extraction of the gabarit.

7.4.3.1 Localization of the External and Internal Edges Our model of localization of the external and internal edges of the iris is based on the geometric determination of a ring delimited by two conics (i.e., circles) C1 and C2 (cf. Chapter 5, Section 5.3) in the binary image in grayscale I_g, where

- C1 is the external edge.
- C2 is the internal edge.

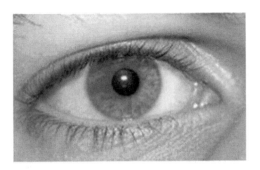

Figure 7.3 Binary image in grayscale level Ig.

These two conics are characterized by the parameters $(\rho_{iris}, c_i(x_i, y_i))$ for C1 and $(\rho_{pupil}, c_p(x_p, y_p))$ for the edge C2, where

- ρ_{iris} and ρ_{pupil}: radii (ρ_{iris} and $\rho_{pupil} \in P$)
- c_i and c_p: centers, respectively, defined by their coordinates (x_i, y_i) and (x_p, y_p) in a Cartesian system

We applied the Canny operator, using the Canny function automatically generated by MATLAB, on the image I_g. This function uses the standard deviation σ of the Gaussian filter (cf. Chapter 2, Section 2.5.1), with $\sigma = 1$. We obtain a binary image I_b, as shown in Figure 7.4, in the form of a matrix composed of 1 (white) and 0 (black). This matrix has a size of 240×320.

We applied the circular Hough transform on the image I_b for the localization of the edges C1 and C2. This method has been adopted by Daouk et al. (2002) and Tian et al. (2004). However, we have integrated the aspects of discrete geometry for defining the presence of an edge (cf. Chapter 2, Section 2.6) in an image. Our method aims to scan only the elements of the image I_b that have a value 1 (i.e., presence of an edge) (cf. Chapter 2, Section 2.6), which reduces the time taken to locate the edges to an average of 5.5 seconds (cf. Chapter 10, Section 10.1.1).

The localization of the edges C1 and C2 occurs in two successive steps:

1. *Localization of the edge* C1 in the image I_b
2. *Localization of the edge* C2 in the image Ic, where Ic represents a rectangular region taken in the image I_b. This region is bounded by the four poles of the edge C1, as shown in Figure 7.5, located in the first step.

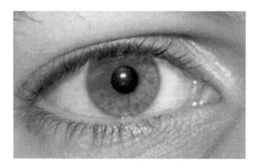

Figure 7.4   Binary image Ib as result of Canny.

Figure 7.5   Image Ic delimited by the edge C1.

The objective of using the image Ic is to reduce the time of localization of the edge C2. The result of the localization of both C1 and C2 is illustrated in Figure 7.6.

7.4.3.2 Normalization of the Iris Region   Our model of normalization of the region of the iris consists first in isolating the region of the iris Is from the image Ib and then in eliminating the effect of the eyelids that mask a part of the iris in order to obtain a raster T.

7.4.3.2.1 Isolation of the Region of the Iris   The isolation of the region of the iris consists in isolating the region of the iris Is, which is delimited by C1 and C2 in the image Ic. The method of isolation of Is corresponds to a transformation of Cartesian coordinates (i.e., ring) to polar coordinates (i.e., rectangle) (cf. Chapter 2, Section 2.5.3). Each point of Is is characterized by its coordinates (x, y) in Cartesian theory and its

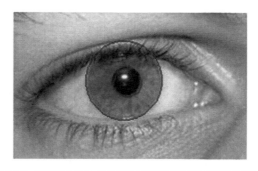

Figure 7.6   Representation of the edges C1 and C2 in the image Ib.

density of gray level $0 \leq Is(x,y) \leq 255$. By passing to polar coordinates, we obtain the following system:

$$0 \leq \theta \leq 2\pi$$

$$\rho_{pupil} \leq \rho \leq \rho_{iris}$$

$$Is(\rho, \theta) = Is(x, y)$$

where ρ_{pupil} is the radius of the pupil, and ρ_{iris} is the radius of the iris.

Figure 7.7 illustrates the representation of the isolated iris region Is in Cartesian coordinates (Figure 7.7a) and in polar coordinates (Figure 7.7b). The latter are represented as a frame raster where the lines are marked by ρ and the columns by θ. Its size is of 25×3601 points (or elements).

7.4.3.2.2 Elimination of the Effects of the Eyelids   For the elimination of the effects of upper and lower eyelids, we used a mask (Chassery and Montanvert 1991) of value −1. This mask eliminates both the upper and lower parts of the isolated iris region Is. It hides the points of Is not belonging to the region of intersection between the isolated iris region Is and an ellipse (defined later) by assigning values of −1. The complement of this hidden part corresponds to the intersection region and having actual value taken from the raster Is. The result after the application of the mask is represented as a raster $T(m,n)$ having the same size as Is, where m and n, respectively, represent the number of rows and columns of this raster.

Note that this ellipse is defined by the following parameters:

- a = large radius = ρ_{iris} + 10 pixels
- b = small radius = ρ_{iris} − 10 pixels

Our choice of a and b is based on a sample of 257 different iris images from the CASIA iris image database (CASIA 2006). These

(a)   (b)

Figure 7.7   Representation of the iris in Cartesian and polar coordinates.

images have different positions of the eyelids. We observed that, on average, the eyelid hides 10 pixels of the radius of the iris ρ_{iris} of both the upper and lower poles. With the elimination of 10 pixels of these two poles, we keep most of the biometric features for verification and we have more efficiency (cf. Chapter 10, Section 10.2.3) in comparison with other methods (cf. Chapter 6, Section 6.2).

The determination of the raster T(m,n) is as follows:

- Initialization of the raster T(m,n) to –1
- *Calculation of the angle* θ', which determines the four points of intersection (M_1, M_2, M_3, and M_4) between the external edge of the iris C1 and the ellipse, as shown in Figure 7.8

$$\theta' = \arctan\left(\left(\frac{b}{a}\right) \times \sqrt{\frac{a^2 - \rho_{iris}^2}{\rho_{iris}^2 - b^2}}\right) \tag{7.1}$$

- Determination of the values of the elements representing the intersection region by dividing Is into four parts (Figure 7.8) and assigning the values of the elements of these parts to the corresponding elements in T, with:

1. Part 1 exists between the two vectors $\overline{OM_1}$ and $\overline{OM_2}$. It corresponds to the angle θ existing between $(0, \theta')$ and $((2\pi - \theta' + 0.1), 2\pi)$ and the radii ρ (in polar coordinates) varying between 1 and $(\rho_{iris} - \rho_{pupil} + 1)$.

2. Part 2 exists between the two vectors $\overline{OM_1}$ and $\overline{OM_3}$. It corresponds to the angles θ ranging from $(\theta' + 0.1)$ to $(\pi - \theta')$ and to the radii ρ (in polar coordinates) ranging between 1 and $(\rho_e - \rho_{pupil} + 1)$, where ρ_e (Postnikov 1981) represents the radius in polar coordinates moving along the ellipse with

$$\rho_e = \frac{(a \times b)}{\sqrt{(b \times \cos(\theta))^2 + (a \times \sin(\theta))^2}} \tag{7.2}$$

3. Part 3 exists between both segments $\overline{OM_3}$ and $\overline{OM_4}$. It corresponds to the angles θ varying between $(\pi - \theta' + 0.1)$ and $(\theta + \pi)$ and in radii ρ ranging between 1 and $(\rho_{iris} - \rho_{pupil} + 1)$.

(a)   (b)

Figure 7.8 Representation of the intersection region between the ellipse and the isolated region of the iris, Is, and the raster, T.

4. Part 4 exists between both segments $\overline{OM_4}$ and $\overline{OM_2}$. It corresponds to the angles θ varying between $(\pi + \theta' + 0.1)$ and $(2\pi - \theta)$ and in radii ρ ranging between 1 and $(\rho_e - \rho_{pupil} + 1)$, where ρ and ρ_e are defined previously.

Figure 7.8 illustrates the representation of the region of intersection between the ellipse and the region of the iris isolated Is (Figure 7.8a), as well as the raster T obtained after elimination of the effects of the upper and lower eyelids (Figure 7.8b).

7.4.3.3 Extraction of Gabarit   Our model of extraction of gabarit consists in extracting the iris biometric features gabarit (cf. Chapter 2, Section 2.5.5) of the raster T. In the following, this gabarit will be represented by the gabarit DHVA and designated by g.

We used the two-dimensional discrete Haar wavelet adopted by Lim et al. (2001) (cf. Chapter 2, Section 2.5.5) for the extraction. This is to keep more features for the verification and to have more effectiveness because the data of the iris are very sensitive.

This Haar wavelet is automatically generated by MATLAB—obtaining, as a result, the approximation coefficient CA and the detail coefficients CH (horizontal coefficient), CV (vertical coefficient), and CD (diagonal coefficient). Each of these coefficients is represented as a matrix of size 11×1801 elements in average in the number of lines. This variation depends on the size of the iris affected by the distance between the camera and the eye, as well as the dilation of the pupil due to the effect of illumination. Figure 7.9 illustrates these representations.

All these matrices are combined together in the order of CD, CH, CV, CA to form a single two-dimensional vector, as shown in Figure 7.10. This vector represents the biometric features (or biometric signature) of the iris gabarit DHVA to be compared to other gabarits

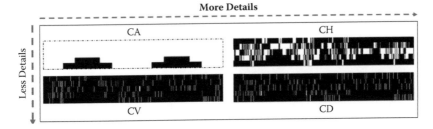

Figure 7.9   Representation of the coefficients CA, CH, CV, and CD resulting from the two-dimensional Haar wavelet.

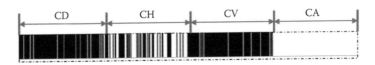

Figure 7.10   Representation of the gabarit DHVA.

DHVA stored in the database, as detailed in Section 7.4.5. The size of this vector is 11×7204 elements. This concept was introduced by Tian et al. (2006) for the verification.

7.4.4 Phase 4—Encryption of the Gabarit

The phase of encryption of the gabarit aims to encrypt the gabarit DHVA required by the method of asymmetric cryptography using the RSA algorithm (Stallings 1999). This gabarit will be compared to those stored in the hashing file (Figure 7.11) by the search method detailed in the following section. This encrypted gabarit will be stored in a secure manner in the hashing file according to the method of classification of gabarits (detailed in the next section) to prevent its easy recovery.

7.4.5 Phase 5—Verification

The verification phase operates in two modes: identification and authentication. The identification of a person, using biometric features of the iris, allows this person access to a computer system. The authentication of a person serves to check the access permissions.

In order to accelerate the verification process, especially for a large database, we propose a method of classification of the gabarits DHVA. This method leads to obtaining a membership to a certain partition in the database according to the criterion "average of gray" relative to each gabarit DHVA.

7.4.5.1 Classification of the Gabarit

7.4.5.1.1 Architecture of the Model Our classification model of the gabarits DHVA issued from our model is based on the indexed hierarchy by trees (cf. Chapter 2, Section 2.7.3). This model is required to meet the following criteria:

- Group the personal information, including the gabarits DHVA, in a structured way.
- Obtain with a fast response time the verification result by directly accessing the partition containing the gabarits DHVA that we want to compare with the required gabarits DHVA.
- Have a maximum yield search, as for an organization of the tree, which reduces the number of disk consultations by avoiding the empty nodes.
- Have a minimum size of the tree and a reduced number of balancing operations.
- Be able to expand the storage space in a dynamic way as with the addition of new gabarits DHVA.
- Have an unlimited number of groups.

Our classification method of the gabarits DHVA is based on two criteria:

1. The percentage of the average of gray of the raster T related to the gabarit DHVA, denoted by g, within a scale of 0 to 100 and rounded to the two nearest decimals. Each raster T is represented as a matrix T(m,n), where
   - m is number of lines.
   - n is number of columns.

$$\overline{Mg} = \left| \frac{\displaystyle\sum_{i=1}^{m}\sum_{j=1}^{n} T(i,j)}{m \times n} \right| \times \left(\frac{100}{255} \right) \tag{7.3}$$

2. The cardinal N is the maximum number of indices (cf. Chapter 2, Section 2.7.3) of the elements that correspond to the percentage of the average of gray (u = 1, ..., N) of the raster related to the gabarits g.

These elements belong to the same subgroup relative to a group p (p = 1, ..., ∞).

The cardinal N is determined by the distribution of the set of $\overline{M}$g:

$$N = \text{card}(\overline{M}\text{gucp}) \tag{7.4}$$

Following the results of the simulations performed on a sample of 257 different iris images issued from the CASIA iris image database (CASIA 2006) (cf. Chapter 10, Section 10.3), we opted for the optimum value of N to be equal to 50.

Our grouping method of such structures, gabarits, g, in p groups is affected in the same way as in the construction of a B-tree (cf. Chapter 2, Section 2.7.3). However, we limited the number of levels (cf. Chapter 2, Section 2.7.3) to two, where

- Level 1 represents the root r containing p groups that are indexed by a threshold t.
- Level 2 shows the leaf nodes containing the nodes Bcp of groups p.

The tree structure evolves in a dynamic way with the addition of new gabarits g.

The index $\overline{M}$gcp of the key Kp for a group, p, at the level of a root, r, represents the integer part of the percentage of the average of gray $\overline{M}$g incremented by one:

$$\overline{M}\text{gcp} = \text{int}(\overline{M}\text{g}) + 1 \tag{7.5}$$

This value is based on the two criteria mentioned previously.

Each group p refers to a subgroup in the hierarchy or a node Bcp in the tree containing indexes $\overline{M}$gucp that are strictly less than the index $\overline{M}$gcp.

Each index $\overline{M}$gucp, in a given node at the level of leaf, points to a hashing file (Mannino 2004) containing the different gabarits g_t (t = 1, ..., ∞) having the same $\overline{M}$gucp. Each gabarit g_t points to a

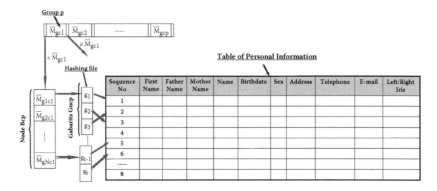

Figure 7.11   Model of classification of the gabarit DHVA based on the percentage of the average of grayscale of rasters.

record in a table containing all the personal information of the concerned person. Note that the order of placement of gabarits g_t does not depend on the order of placement of corresponding personal information. Figure 7.11 illustrates these representations.

7.4.5.1.2 Method of Splitting Node in Case of Supersaturating   When the node Bcp becomes supersaturated (i.e., card $(\overline{M}gucp) = N$), for each arrival of a new index $\overline{M}gucp$, we split the node into two nodes (Bcp1, Bcp2). We move up one level the index in the middle $\overline{M}gucp$, which is then inserted into the parent node, "the root," as a new key Kp (cf. Chapter 2, Section 2.7.3). This key points to two nodes: Bcp1 from the left and Bcp2 from the right side. Figure 7.12 illustrates the process of splitting node BCP into two nodes, Bcp1 and Bcp2, by fixing the cardinal N to 4.

7.4.5.1.3 Method of Grouping of Gabarits DHVA   From the pretopological point of view, this model uses induced pretopological structures (cf. Chapter 2, Section 2.7.2) the gabarits DHVA. These structures are of type ς_S (cf. Chapter 2, Section 2.7.1) defined on a not-empty set E where the number of elements is not limited.

The grouping method of these structures g into p groups is based on a binary relationship P (cf. Chapter 2, Section 2.7.1). This binary relation is based on a threshold t representing, at the level of the root r, the index $\overline{M}gcp$ of the key Kp for a given group p. It combines two elements in the same group: gabarits DHVA g1 and g2 of E

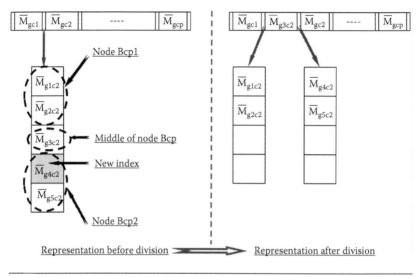

Figure 7.12   Division of node.

corresponding to the percentages of the average of gray $\overline{M}g1cp$ and $\overline{M}g2cp$ strictly less than t, defined by

$$\forall\, g_1\,\varepsilon\, E \quad \forall\, g_2\,\varepsilon\, E$$
$$g_1 \text{ "Pt-similar to" } g_2 \text{ if } \max(\overline{M}g1cp, \overline{M}g2cp) < t$$

Then, for each percentage of the average of gray or greater than or equal to t, they will be distinguished and belong to two different subgroups, with

$$\forall\, g_1\,\varepsilon\, E \quad \forall\, g_2\,\varepsilon\, E$$
$$g_1 \text{ "Pt-dissimilar to" } g_2 \text{ if } \max(\overline{M}g1cp, \overline{M}g2cp) \geq t$$

Figure 7.13 illustrates these presentations to t = 28.

7.4.5.1.4 Method of Search of a Given Gabarit DHVA   The research method of a given structure g_d, in the set of elements E, is based on the concept of search in a B-tree (cf. Chapter 2, Section 2.7.3). This method consists of calculating the percentage of the average of gray, denoted by $\overline{M}gdcp$, for g_d and then looking in the group p indexed by t = int($\overline{M}gdcp$) + 1. However, if the p does not exist or if the structure g_d is not found in p, we adopt the basic concept of neighborhood of base (cf. Chapter 2, Section 2.7.1) of g_d denoted by $\varsigma(g_d)$.

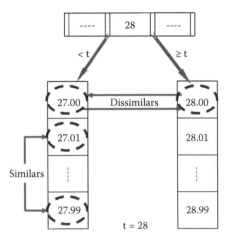

Figure 7.13   Grouping based on binary relation P.

We define the set of structures g belonging to groups p indexed by (t − 2, t − 1, t + 1, and t +2) as the base of neighborhood of g_d indexed by t. The structures related to groups indexed by t − 2 and t − 1 are considered more refined structures (cf. Chapter 2, Section 2.7.1) than those relating to t; the structures related to groups indexed by t + 1 and t + 2 are considered less refined structures (cf. Chapter 2, Section 2.7.1) than those relating to t.

The research method in the groups constituting $\varsigma(g_d)$ is recursive. Our approach aims to look at all elements, respectively, related to groups indexed by t − 1, t + 1, t − 2, and t + 2 to find the "nearest" element (cf. Chapter 2, Section 2.7.1) of g_d (in pretopology). Figure 7.14 illustrates the representations of relationship of refinement "finesse" of all structures, function of t, defined on the set E.

Our model of identification aims to check whether the required personal information exists in the database "personal information" (Figure 7.11). If this information is already stored, then the person is identified before accessing the computer system. Otherwise, we store this information after obtaining the encrypted gabarit DHVA. This last one will be saved in the hashing file according to the classification method of the gabarits DHVA and will point to this information.

7.4.5.2 Authentication   Our model of authentication allows verifying first if the required personal information exists in the database

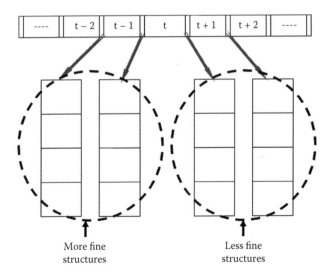

	t − 2	t − 1	t	t + 1	t + 2	
----						----

More fine
structures

Less fine
structures

Figure 7.14   Representation of the relations of fineness.

"personal information," as shown in Figure 7.11. If this information is already stored, in the second step we verify if the required encrypted gabarit DHVA is already stored in the hashing file, as shown in Figure 7.11. For the consultation, we use the search method detailed in Section 7.4.5.1.

Our approach of verification of this gabarit is first to decrypt it using the RSA algorithm (Stallings 1999) and then compare it with the gabarit DHVA stored in the hashing file decrypted with the same algorithm.

The comparison method of the required gabarit DHVA $g_1(m_1,n_1)$ with the gabarit DHVA stored in the database $g_2(m_2,n_2)$ is based on the concept of comparing two vectors adopted by Tian et al. (2006) (cf. Chapter 2, Section 2.5.5), where

- m_1 and m_2: number of lines of g_1 and g_2
- n_1 and n_2: number of lines of g_1 and g_2

However, we compared these two vectors element by element to determine the number of dissimilar elements between these two vectors, which is represented by $d(g_1,g_2)$. Subsequently, we calculated the rate of dissimilarity (cf. Chapter 6, Section 6.2) between these two vectors, designated by rate (g_1,g_2), with

$$\text{rate}(g_1, g_2) = \frac{d(g_1, g_2)}{\alpha \times \beta} \tag{7.6}$$

where $\alpha = \text{Min}(m_1, m_2)$ and $\beta = \text{Min}(n_1, n_2)$.

Note that we have generalized the formula of dissimilarity rate between g_1 and g_2 in this way to consider the case where these two gabarits are not the same size. This difference is due to the factor of distance between the eye and the camera, as well as the illumination factor, which dilates the pupil (cf. Section 7.4.3.3).

Recall that for the calculation of the number of dissimilar elements between g_1 and g_2, we have generalized the range of variation of the number of lines between $i_1 = \alpha - \beta + 1$ and $i_2 = \alpha$ and the range of variation of the number of columns between $j_1 = 1$ and $j_2 = \beta$.

We adopted a threshold value of 0.0200 to decide the equality of these two vectors. In other words,

- If $\text{rate}(g_1, g_2) = 0$, then the two vectors are absolutely identical.[*]
- If $\text{rate}(g_1, g_2) \le 0.0200$, then the two vectors are considered equal.[†]
- If $\text{rate}(g_1, g_2) > 0.0200$, then the two vectors are different.[‡]

The threshold value is estimated using the simulations performed on a sample of 257 different iris images issued from the CASIA iris image database (CASIA 2006) (cf. Chapter 10, Section 10.2.4).

7.5 Conclusion

In this chapter we have developed our biometric model for authentication of users accessing a computer system based on biometric features of the iris of the eye gabarit DHVA. This model ensures secure access to confidential information (i.e., stored in the database or transmitted through the network).

Our model of localization of external and internal edges led to a reduction of the computation time to 5.5 seconds on average.

---

[*] The person is accepted by the system.

[†] This is the case of the same person with the same iris. This person is accepted also by the system.

[‡] This is the case of two different persons or the case of the left iris and the right iris of the same person. In this case, the person will be rejected by the system.

We have introduced a novel method to eliminate the effects of the eyelids, while keeping the essential biometric features of the iris for identification and authentication of people.

By introducing pretopoligical aspects, our classification model of gabarits based on the concept of indexed hierarchical classification allows one to speed the access and the search of gabarits DHVA better than the sequential method. This biometric model should be incorporated into different systems for the authentication of individuals accessing confidential information.

In the next chapter, we present the architecture of our global model, including biometric and cryptographic aspects, integrating a multiagent system.

8

GLOBAL MODEL INTEGRATING A MULTIAGENT SYSTEM

This chapter presents the global model. This model, whose architecture is described in this chapter, is based on the analysis of iris images for authentication of users accessing a system, as well as the transmission of encrypted information, by the asymmetric cryptography technique, through the Internet. In this model, we propose to integrate a multiagent system (MAS) to treat the complexity problems in an organized way in the management of the operation of the system. The architecture of the MAS and the detailed description of the interactions between agents constituting this model will also be presented.

8.1 Introduction

The global model described in the following section aims to secure access to data and their transmission through the Internet. The contribution of a multiagent architecture puts emphasis on the parallelism, the focusing ability, the heterogeneous problem solving, and the reliability (Crevier and Lepage 1997). Our reflection is based on the integration of MAS in our global model. These systems meet the multiexpertise necessary for the model.

8.2 Global Model

Our model is based on the biometric method using the iris of the eye for the authentication of users, and the method of asymmetric cryptography using the Rivest–Shamir–Adleman (RSA) algorithm (Stallings 1999) for the encryption of data. This model should ensure secure access to information and guarantee the protection of confidential information, especially that transmitted through the Internet.

8.2.1 Architecture of the Model

The architecture of the model is composed of the following:

- A user U_A and a user U_B, that want to exchange encrypted information through the Internet, are installed on both ends A and B. These users play the role of a sender or a receiver. We consider that the user U_A is a sender and the user U_B is a receiver.
- Two machines of iris scan, for the authentication of users accessing the system, are installed on both side ends of A and B.
- Two local servers, A and B, are installed on both sides and have two local databases (DB). Both databases contain encrypted private keys of users and encrypted private keys to decrypt the public key of the receiver.
- Two firewalls are also installed at both local servers.
- An authentication server with a public database contains the encrypted public keys of all users and has high safety standards.

In this model, the user U_A and the user U_B each has its own private and public keys. Private keys are stored in the local database, and public keys are stored in the public database.

On both ends A and B we have installed cameras to take suitable images of the iris of the eye. In our experiments we used the Sony CCD color camera with a resolution of 520 TV lines. We applied our biometric model (cf. Chapter 7, Section 7.3) on the captured image, to obtain the iris biometric features gabarit diagonal horizontal vertical approximation (DHVA) (cf. Chapter 7, Section 7.4.3.3).

Each user accessing the system has a composite identifier consisting of personal information (cf. Chapter 7, Section 7.3) and a password based on established coding from the gabarit DHVA. In the remainder of this chapter, the gabarit g will be considered type gabarit DHVA. For data transmitted through the network we have adopted the method of asymmetric cryptography using the RSA algorithm (Stallings 1999). Firewalls are used to filter incoming and outgoing requests with different external addresses internally to deal with external attacks. These firewalls require

- Two types of input ports:
  - One for receiving encrypted messages, coming from the outside, defined by the type of application

- One for receiving encrypted public keys of concerned receivers searched by the sender
- Two types of output ports:
  - One for transmitting encrypted messages as defined by the type of application
  - One for transmission of a request for requesting the public key of the relevant receiver

Both local servers and the authentication server are equipped with antispam and antivirus software to avoid the risk of threats, attacks, infection, and destruction. These applications should be updated regularly to protect the system against any danger caused by viruses and intrusions. Figure 8.1 shows the overall architecture of the system.

If the user U_A (sender) wants to send an encrypted message (M) to the user U_B (receiver) through the network, the process involves the following steps:

- The user U_A should be authenticated by his personal information and his password, composed of biometric features of the iris of his eye "gabarit g," to validate the relay.
- The system S_A captures data related to the iris of the user U_A, encodes them, and then encrypts them to obtain the encrypted gabarit g.

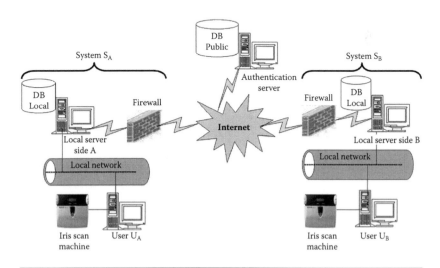

Figure 8.1 Architecture of global model.

- After verification of personal information, the system S_A compares the encrypted gabarit g with those stored in the local database to allow the user U_A to access the system S_A.
- The user U_A sends a request through the local server to fetch the public key of the user U_B stored on the authentication server in encrypted form. The authentication server responds to the request of the user U_A after asking for confirmation from the user U_B.
- The user U_A decrypts the public key of the user U_B for use in encrypting the clear message by applying the RSA cryptography algorithm (Stallings 1999).
- The system S_A prepares the message, encrypts it, and sends it to the system S_B. The encrypted message M passes through a port defined by the firewall installed on the local server side S_A. Then, it is received by the local server side S_B through a port defined by the firewall installed at this server.
- The user U_B will access the encrypted message.
- The user U_B authenticates and then decrypts the encrypted message with the same RSA algorithm and consults it.

8.2.2 Representation of Interactions between Different Actors

The interactions between the different actors of the conceived global model are described in the sequence diagram shown in Figure 8.2. The notations used in the figure are as follows:

[HR]:	[having **res**ponse]
[BM]:	[**b**lock **m**essage]
[BEM]:	[**b**lock **e**ncrypted **m**essage]
[BR]:	[**b**lock **r**equest]
[FPK]:	[(if response HR is positive, **f**ind the **p**rivate **k**ey of receiver to decrypt the encrypted message, access nonauthorized)]
[FPKR]:	[(if connected sender, **f**ind the **p**ublic **k**ey of **re**ceiver, access nonauthorized)]

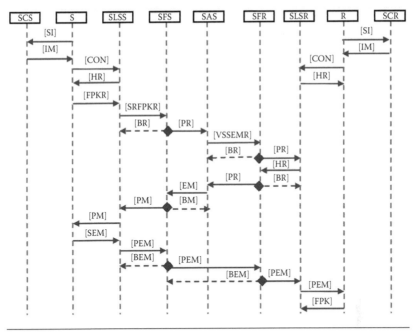

Figure 8.2   Interactions between the different actors of the global model.

[CON]:	[**con**nect (personnel information, password = gabarit g)]
S:	sender
[SM]:	[(if response positive, **s**end **m**essage contains the encrypted public key of receiver, **s**end an empty **m**essage)]
[SEM]:	[(is reception of public key of receiver, **s**end the encrypted **m**essage, stop)]
[SRFPKR]:	[**s**end **r**equest to **f**ind the **p**ublic **k**ey of **r**eceiver]
[IM]:	[having an **i**ssued **m**odel = gabarit g]
[PM]:	[**p**ass **m**essage]
[PEM]:	[**p**ass **e**ncrypted **m**essage]
[PR]:	[**p**ass **r**equest]
R:	receiver
SCS:	system with **c**amera **s**ender side
SCR:	system with **c**amera **r**eceiver side
[SI]:	[**s**can the **i**ris]
SFS:	system **f**irewall **s**ender side
SFR:	system **f**irewall **r**eceiver side

SAS: system authentication server
SLSS: system local server sender side
SLSR: system local server receiver side
[VSSEMR]: [verify if sender can send encrypted messages
                  to receiver]

8.2.3 Composition of the Model

Our global model is composed of two modules:

- Module 1—biometric authentication using the iris of the eye
- Module 2—transmission of encrypted information through the Internet

These two modules are each represented by a flow diagram (Mannino 2004) showing the different information flows between the various processes to achieve a specific goal (e.g., authenticated person, identified person, or encrypted sent message).

In the following, we consider that the private key and the public key are not the same as the private key and the public key of the receiver. They are common to all users and stored locally on the user's machine.

8.2.3.1 Module 1—Biometric Authentication Using the Iris of the Eye   Module 1 is represented by a flow diagram, as shown in Figure 8.3, containing all the processes and the flow of information to validate users' access to the system according to biometric features of the iris eye gabarit g. This diagram describes the different processes of the identification mode (cf. Chapter 7, Section 7.4.5.2) and the various processes of the authentication mode (cf. Chapter 7, Section 7.4.5.3). Both methods have processes in common as well as different processes.

The set of these key processes B_i ($i = 1, \ldots, 12$) is established according to a sequential logic and sometimes parallel as follows:

1. Process B_1 (1.1, 1.2): enter the personal information:
   a. In the case of identification, if the personal information already exists, the person is rejected and the new person will be informed.

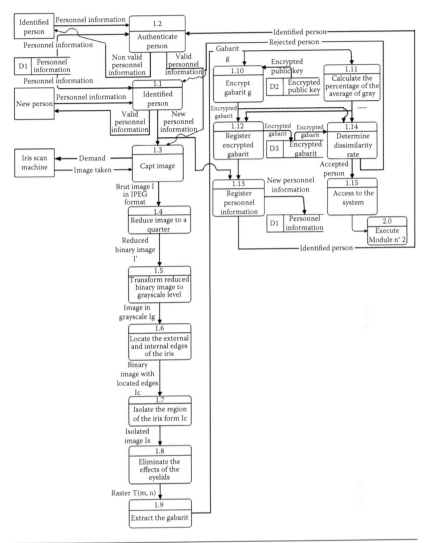

Figure 8.3 Representation of module 1—biometric authentication using the iris of the eye.

b. In the case of authentication, if the personal information is not valid, the person is rejected and the person already identified will be informed with the possibility to reenter the personal information.

c. In case the personal information is new for identification or valid authentication, perform the nine following processes B_k (k = 2, ..., 10).

2. Process B_2 (1.3): capture the gross image I in JPEG format.

3. Process B_3 (1.4): reduce the image I to a quarter to obtain a reduced binary image I'.

4. Process B_4 (1.5): transform I' to grayscale to obtain the image in grayscale Ig.

5. Process B_5 (1.6): locate the two external and internal edges of the iris to obtain a binary image with located edges Ic.

6. Process B_6 (1.7): isolate the region of the iris from Ic to obtain an isolated image Is.

7. Process B_7 (1.8): eliminate the effect of the eyelids to obtain a raster $T(m,n)$.

8. Process B_8 (1.9): extract the biometric features of the iris of the eye to obtain a gabarit g.

9. Process B_9 (1.10): encrypt the gabarit g to obtain an encrypted gabarit g.

10. Process B_{10} (1.11): calculate the percentage of the average of gray $\overline{Mg}$ (cf. Chapter 7, Section 7.4.5.1) relative to the gabarit g.

11. Process B_{11} (1.12, 1.13): in case of identification and for the person to be identified and able to be authenticated:
    a. Record the encrypted gabarit g.
    b. Record the personal information.

12. Process B_{12} (1.14, 1.15, 2.0): in case of authentication:
    a. Determine the rate of dissimilarity between the required gabarit g and those stored in the database to decide about the conformity.
    b. In case of conformity, the user is granted access to the system and is ready to execute module 2.
    c. In case of nonconformity, the user is rejected by the system and has the possibility to be identified again.

8.2.3.2 Module 2—Transmission of Encrypted Information across the Internet Module 2 is represented by a flow diagram, as shown in Figure 8.4. This diagram contains all the processes and the flows of information describing the preparation of a clear message from the user U_A side (sender), its transmission as an encrypted message M through the Internet, and its consultation by the user U_B (receiver) after decryption.

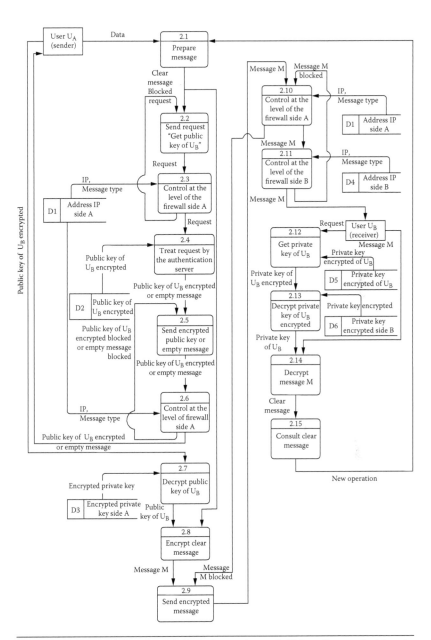

Figure 8.4 Representation of module 2—exchange of encrypted information over the Internet.

The set of these main processes R_i $(i = 1, \ldots, 12)$ is established according to a sequential logic for each of the two sides (A and B) as follows:

1. Process R_1 (2.1): prepare the clear message by the user U_A as sender.
2. Process R_2 (2.2): from the user U_A side, send a search query of the public key of the user U_B as a receiver.
3. Process R_3 (2.3, 2.4): check this request at the level of the firewall side a:
    a. In case of nonvalid requests, a blocked message is returned with possibility to launch the request again.
    b. In cases of valid requests, treat the request at the level of the authentication server.
4. Process R_4 (2.5): verify at the level of the authentication if user U_A can send encrypted messages to user U_B:
    a. In cases when user U_A cannot send encrypted messages to user U_B, send a blank message to the user U_A.
    b. If not, send a message that contains the encrypted public key of the user U_B.
5. Process R_5 (2.6): control at the level of the firewall side A of the sent messages from the authentication server:
    a. In case of blocking, send the messages again.
    b. If not, send messages to the user U_A.
6. Process R_6 (2.7, 2.8, 2.9): in case of the user U_A receive a message containing the encrypted public key of the user U_B:
    a. Decrypt the encrypted public key of the user U_B.
    b. Encrypt the clear message to obtain an encrypted message M.
    c. Send the encrypted message M to the user U_B by passing by the firewall side A and the firewall side B.
7. Process R_7 (2.10): control the encrypted message M at the level of the firewall side A:
    a. In case of blocking, send the encrypted message M again.
    b. If not, execute the following process.
8. Process R_8 (2.11): control the encrypted message M at the level of the firewall side B:
    a. In case of blocking, return to process R_7.
    b. If not, execute the following process.

9. Process R_9 (2.12): find the private key of the user U_B to decrypt the encrypted message M.
10. Process R_{10} (2.13): decrypt the private key of the user U_B.
11. Process R_{11} (2.14): decrypt the message M using the private key of the user U_B to obtain the clear message.
12. Process R_{12} (2.15): consult the clear message and be ready for a new operation.

8.3 Integration of Multiagent System

We were brought to conceptualize an MAS to model rationally throughout our system. The MAS system is composed of specialized agents involved at various levels such as the authentication of users accessing the system, the analysis of the biometric features of the iris, the encryption, and the data transmission through the Internet.

The MAS is heterogeneous because it is composed of agents from different levels of intelligence. Some agents are reactive and others have a certain level of intelligence allowing them to perceive their environment (i.e., estimate of the distance between the person and the camera) and take decisions based on the state of the environment.

8.3.1 General Model of Integrated MAS

Each agent should perform a task based on a predefined plan. At each step of the plan, an agent should possess only the tools needed to carry out its work. Each agent has a set of tasks to accomplish its plan. The agents cooperate to achieve their goals by exchanging messages. Our communication protocol is of question/response type in a client/server model. The relationships between the activities of agents and the information needed to carry out these activities follow a sequential or parallel logic.

The communication between agents of our system is done by pools (families of stable data can be created, viewed, modified, or destroyed) or by channels (families of data of type "first in, first out" [FIFO], which are produced to be consumed). This system includes six types of agents:

1. The agent system user (ASU) manages the treatment of iris image for authentication of users accessing the system.
2. The agent user (AU) manages the estimation of parameters passed to the ASU. As well, it manages the emission and the reception of an encrypted message through the Internet.
3. The agent firewall (AF) monitors the incoming and outgoing requests to and from the system to counter any attack.
4. The agent authentication server (AAS) monitors the search for and the sending of the public key of the receiver to encrypt the clear message prepared by the sender.
5. The agent key (AK) controls the search of private keys and public keys of the users of the system.
6. The agent verification (AV) manages the identification and the authentication of a person who wants to access the system.

In the following, we consider the active instances of side A (sender) and those on the side B (receiver) that we will continue to call agents.

8.3.2 Architecture of Agents

Each agent of our MAS is executing a set of tasks in a sequential or parallel way. Each agent has internal management and relations with its environment. The agent represented by one or more tasks includes an initialization phase preparing the agent to carry out its work. This phase is performed only once. Initialized data in this phase are valid throughout the life of the agent. The treatment processes at the level of each agent are illustrated in Figure 8.5.

The procedure performed by the agent is composed of an endless loop including the following steps:

- The agent waits for a message. When a message is received, it will be classified according to its content and its importance in the queue.
- The agent should parse the received message for processing by appropriate procedures or to request services by sending messages to other agents.
- The agent receives the following message in the order of priority of treatment or waits if no receiving message enabling it to pursue its plan.

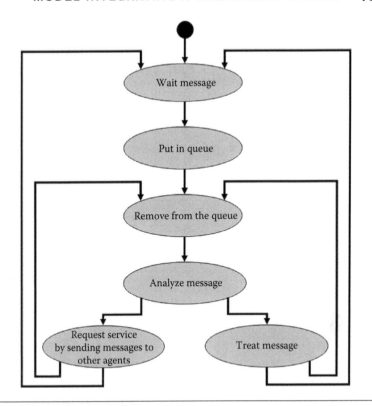

Figure 8.5 Agent treatments diagram.

- If the requested services by the agent are not available, the agent tries to continue executing from its plan, in a separate phase from what it has not done.
- In case the agent cannot receive the requested services or continue with its plan to another phase, the agent waits.

The received messages by the agent are considered application messages, forming part of the plan to be performed by the agent.

8.3.3 Realization of Instances of MAS Agents

In this section we describe the behavior of different instances of agents integrated in the global model. Each of these instances has several tasks based on the objectives set out by the system. In the following, we consider the active instances of side A (sender) and those on the side B (receiver) that we will continue to call agents. We also denote the following in the rest of this chapter:

- The ASU represents the agent system user side A (ASU_A) or the agent system user side B (ASU_B).
- The AU represents the agent user side A (AU_A) or the agent user side B (AU_B).
- The AK represents the agent key side A (AK_A) or the agent key side B (AK_B).
- The AV represents the agent verification side A (AV_A) or the agent verification side B (AV_B).
- The AF is the agent firewall side A (AF_A) or the agent firewall side B (AF_B).
- The AAS is the agent authentication server.
- The system user represents the machine installed on the ends A or B. It is by this machine that the user will validate his access and perform his work.

8.3.3.1 Agent System User   The ASU is located at sides A and B of the user systems. It enters the personal information of the examined user. It captures the gross image containing the iris of the eye of this user. It converts this image into a grayscale image (Ig).

The ASU asks AU by giving it Ig to estimate the parameters for locating both external and internal edges of the iris. The AU determines the lower limit ri1 and the upper limit ri2 of the interval [ri1,ri2] of the search of the radius ρ_{iris} of the external edge C1 of the iris. Similarly, the lower limit rp1 and the upper limit rp2 of the interval [rp1,rp2] of the search of the radius ρ_{pupil} of the internal edge C2 of the iris are also determined by this agent AU. Figure 8.6 illustrates these presentations.

The ASU extracts the gabarit g and encrypts it by a public key. This public key was sent by the AK passing through the AU. It performs the calculation of the percentage of the average of gray $\overline{Mg}$ on each gabarit g. Figure 8.7 shows the class of the agent system user.

8.3.3.2 Agent User   The AU is located at both sides A and B of the user systems. It interprets the image Ig received from ASU and estimates the distance d between the user and the camera. Based on this distance, the AU uses its base of rules and makes its calculations to determine the value of the parameters ri1, ri2, rp1, and rp2. It sends

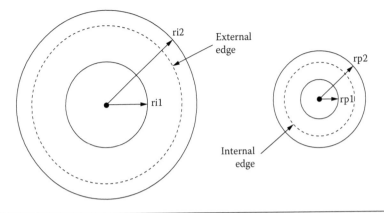

Figure 8.6 Determination of both external and internal edges of the iris.

ASU
−Personnel information −Raw image −Image in grayscale lg −Parameters (ri1, ri2, rp1, rp2) −Gabarit g −Public key −Gabarit g encrypted −Percentage of the average of the grayscale level of the gabarit g
+Add_Personnel_Information() +Capture_Raw_Image() +Transform_Raw_Image() +Find_Parameters() +Extract_Gabarit () +Calculate_Percentage_Average_Grayscale_Level()

Figure 8.7 Class agent system user.

these estimated parameters to the ASU. It asks the AK to find the public key for ASU. After receiving from ASU the encrypted gabarit g, the percentage of the average of gray, and the personal information, the agent interacts with the AV. It asks the AV to check the data relating to the user in question. Figure 8.8 shows the class of the agent user.

8.3.3.3 Agent User Side A The agent user (AU_A) is located at the user system side A. It interacts with the AF_A to seek the public key of the receiver in the AAS. It asks the AK_A the private key to decrypt the public key of the receiver. It encrypts the clear message

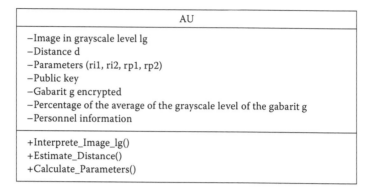

Figure 8.8 Class agent user.

AU_A

−Public key of receiver
−Private key
−Clear message
−Encrypted message M

+Find_Public_Key_Receiver()
+Find_Private_Key()
+Decrypt_Public_Key()
+Encrypt_Clear_Message()
+Send_Encrypted_Message()
+Adjust_Behavior()

Figure 8.9 Class agent user side A.

in an encrypted message M. It also interacts with the AF_A to send the message M to AU_B. In case of warning of blocking, it reacts to see the cause of this blocking. It perceives its environment to adjust its behavior. It decides to send the request again. Figure 8.9 shows the class of the user agent side A.

8.3.3.4 Agent User Side B The agent user (AU_B) is located at the user system side B. It takes the encrypted message M from the AF_B. It interacts with the AK_B to find the private key of the receiver to decrypt the message M. It also interacts with the AK_B to find the private key to decrypt the private key of the receiver. It takes the encrypted private key of the AK_B. It decrypts the private key of the receiver and it decrypts the encrypted message M. Figure 8.10 shows the class of the agent user side B.

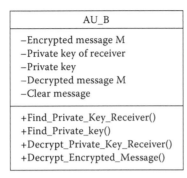

Figure 8.10 Class agent user side B.

8.3.3.5 Agent Firewall Side A The agent firewall (AF_A) is located at the firewall side A. It monitors the incoming and the outgoing requests. It interacts with the AU_A to verify the request of retrieving the public key of the receiver and the encrypted message M sent to the AU_B. It also interacts with the AAS to send the message of search of the public key of the receiver. It also checks the message received from the AAS. Figure 8.11 shows the class of the agent firewall side A.

8.3.3.6 Agent Firewall Side B The agent firewall (AF_B) is located at the firewall side B. It checks the received message from AF_A containing the encrypted message M. It interacts with the AU_B to send the encrypted message M. Figure 8.12 shows the class of the agent firewall side B.

8.3.3.7 Agent Authentication Server The AAS is located at the authentication server. It takes the message containing the request of the public key of the receiver from the AF_A. It reacts autonomously, using

```
+-----------------------------------------------------+
|                        AF_A                         |
+-----------------------------------------------------+
| −Public key of receiver                             |
| −Encrypted message M                                |
| −Message of ASA                                     |
+-----------------------------------------------------+
| +Verify_Request_Public_Key_Receiver()               |
| +Verify_Encrypted_Message()                         |
| +Send_Message_Request_Public_Key_Receiver()         |
+-----------------------------------------------------+
```

Figure 8.11 Class agent firewall side A.

AF_B
−Encrypted message M
+Verify_Encrypted_Message() +Send_Encrypted_Message()

Figure 8.12   Class agent firewall side B.

its base of rules, to decide whether the sender can send encrypted messages to the receiver. In case the sender can send encrypted messages to the receiver, the AAS sends the public key of the receiver in encrypted form to the AU_A by passing through the AF_A. Otherwise, it sends a blank message. In the case of warning of blocking by the AF_A, it reacts to see the cause of this blocking. It perceives its environment to adjust its behavior. It decides to send the message back to AF_A. Figure 8.13 illustrates the class of agent authentication server.

8.3.3.8 Agent Key Side A and Side B   The agent keys (AK_A) and (AK_B) are located at the local servers on both sides A and B. They take from the AU the requests of search of the public key to encrypt the gabarit g. They handle this request and send this key in encrypted form to the AU. The AK_A takes from the AU_A the request of the private key of the receiver to decrypt the encrypted message M. It handles this request and sends this key in encrypted form to the AU_A. Figure 8.14 shows the class of the agent key.

8.3.3.9 Agent Verification Side A and Side B   The agent verifications (AV_A) and (AV_B) are located at the local servers on both sides A and B. These agents manage the identification and authentication of

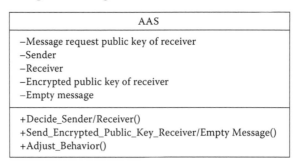

AAS
−Message request public key of receiver −Sender −Receiver −Encrypted public key of receiver −Empty message
+Decide_Sender/Receiver() +Send_Encrypted_Public_Key_Receiver/Empty Message() +Adjust_Behavior()

Figure 8.13   Class agent authentication server.

AK
−Request of public key −Public key encrypted −Request of receiver private key −Encrypted private key of receiver
+Send_Encrypted_Public_Key() +Send_Encrypted_Private_Key_Receiver()

Figure 8.14 Class agent keys.

a user accessing the system. They take from the AU the encrypted gabarit g, the percentage of the average of gray $\overline{\text{Mg}}$, and the personal information. At the level of identification of a person, both agents react autonomously to record the encrypted gabarit g and the personal information. At the authentication of a user, these two agents react autonomously to determine the rate of dissimilarity between the gabarit g required and those stored in the hashing file to make a decision on the conformity and the validation of the system access. Figure 8.15 illustrates the class of the agent verification.

The class diagram in Figure 8.16 represents the various classes of agents and their relationships. These relationships are of type (m, n), since each agent communicates with the neighborhood agents several times to request one or more services by sending messages. The class AU represents the user agent able to authenticate, to receive, and to send messages. The class AF aims to play the role of an agent firewall receiver or sender.

To simplify and clarify the process agent, we have chosen to separate instances "sender" and "receiver" on the user and the agent fire-

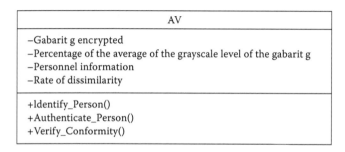

AV
−Gabarit g encrypted −Percentage of the average of the grayscale level of the gabarit g −Personnel information −Rate of dissimilarity
+Identify_Person() +Authenticate_Person() +Verify_Conformity()

Figure 8.15 Class agent verification.

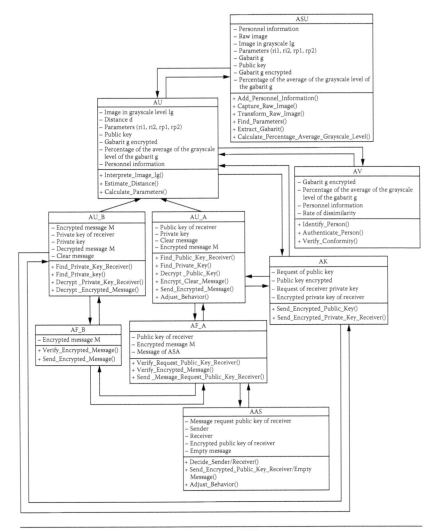

Figure 8.16   Class diagram of agents of our model MAS.

wall. This justifies the presence of classes AU_A, AU_B, AF_A, and AF_B.

$$f(x) = a_0 + \sum_{n=1}^{\infty} \left(a_n \cos \frac{n\pi x}{L} + b_n \sin \frac{n\pi x}{L} \right)$$

8.3.4 Modeling Interactions between Agents

The representation of all tasks allocated to the different agents of our model MAS is based on the concept of distributed planning (cf. Chapter 4, Section 4.6). We represent a flow chart using a set of

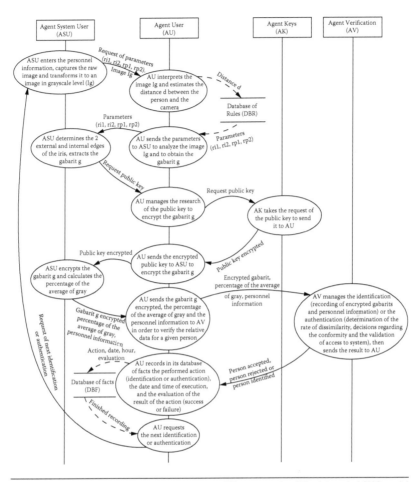

Figure 8.17 Model of interactions between the different agents of the integrated MAS model concerning the biometric authentication.

processes that constitutes our plan. The time of execution of this plan will be optimized by synchronization techniques (Padiou and Sayah 1990) on the tasks performed by the agents.

8.3.4.1 Interactions of Agents for Biometric Authentication The flow diagram illustrated in Figure 8.17 describes all the interactions between the different agents integrated in the global model for biometric authentication using the iris of the eye. The sequence of main processes AB_i ($i = 1, \ldots, 14$) constituting the flow diagram is represented as follows:

1. Process AB_1: the ASU enters the personal information. It captures the gross image containing the iris of the eye and

transforms it to grayscale. Then, it asks AU to get the parameters ri1, ri2, rp1, and rp2 by passing the image in grayscale Ig.

2. Process AB_2: the AU interprets the image Ig and estimates the distance d between the user and the camera.

3. Process AB_3: the AU using the base of rules on which it made its calculations to decide on the value of the parameters ri1, ri2, rp1, and rp2.

4. Process AB_4: the AU sends the parameters ri1, ri2, rp1, and rp2 to the ASU.

5. Process AB_5: the ASU takes the parameters ri1, ri2, rp1, and rp2 to determine both external and internal edges of the iris in the image Ig. It extracts the gabarit g and then requests the AU to fetch the public key to encrypt the gabarit g.

6. Process AB_6: the AU manages the search of the public key to encrypt the gabarit g by communicating with the AK.

7. Process AB_7: the AC takes the request of the public key to send it back to the AU.

8. Process AB_8: the AU sends the encrypted public key to the ASU to encrypt the gabarit g.

9. Process AB_9: the ASU receives from the AU the encrypted public key. It encrypts the gabarit g and calculates the percentage of the average of the gray.

10. Process AB_{10}: the AU receives from the ASU the encrypted gabarit g, the percentage of the average of gray, and the personal information. It sends them to the AV to verify the data relating to the user in question.

11. Process AB_{11}: the AV manages the identification (recording of encrypted gabarits g and of personal information) and the authentication (determining the rate of dissimilarity, decision making regarding the conformity, and validation of the access to the system) of the user in question.

12. Process AB_{12}: the AV sends the result to the AU.

13. Process AB_{13}: the AU records in the base of facts the performed action (identification or authentication) and the date and time of the execution of this action, as well as the evaluation of the result of the action (success or failure).

14. Process AB_{14}: the AU asks for a next identification or authentication.

8.3.4.2 Interactions of Agents for Transmission of Encrypted Data across the Internet   The flow diagram, illustrated in Figure 8.18(a) and (b), describes all the interactions between the different agents integrated in the global model for the transmission of encrypted data over the Internet, as well as between the agent and its environment. The sequence of main processes AR_i ($i = 1, \ldots, 15$) constituting the flow diagram is represented as follows:

1. Process AR_1: the AU_A asks the AF_A to retrieve the public key of the receiver from the AAS.
2. Process AR_2: the AF_A verifies the request of the public key of the receiver to be sent back to the AAS:
   a. In case the request is not valid (i.e., the type of the request sent via the IP address is not defined at the level of the outgoing port of the firewall side A), the AF_A informs the AU_A that the request is blocked. The AU_A reacts to see the cause of blocking. It perceives its environment to adjust its behavior and decides to send the request again.
   b. If not, the AU_A sends this request to the AAS and executes the following process.
3. Process AR_3: the AAS takes the request and, using its base of rules BR_AAS, decides if the sender can send the encrypted messages to the receiver:
   a. In case the sender can send encrypted messages to the receiver, the AAS sends to the AF_A a message containing the encrypted public key of the receiver.
   b. If not, the AAS sends to the AF_A a blank message.
4. Process AR_4: the AF_A manages the verification of the received message from the AAS:
   a. In case the message is not valid (i.e., the type of encrypted message sent via the IP address is not defined at the level of the incoming port of the firewall side A), the AF_A informs the AAS that the message is blocked. The AAS reacts to see the cause of blocking. It perceives its environment to adjust its behavior and decides to send the message again.
   b. If not, the AF_A sends the message to the AU_A and executes the next process.

5. Process AR_5: the AU_A takes the message from the AF_A:
   a. In case of a blank message, the process ends.
   b. If not, execute the next process.
6. Process AR_6: the AU_A asks to retrieve the private key, to decrypt the public key of the receiver, from the AK_A.
7. Process AR_7: the AK_A takes this request and sends the private key in encrypted form to the AU_A.
8. Process AR_8: the AU_A takes the encrypted private key and decrypts it. It encrypts the clear message to encrypted

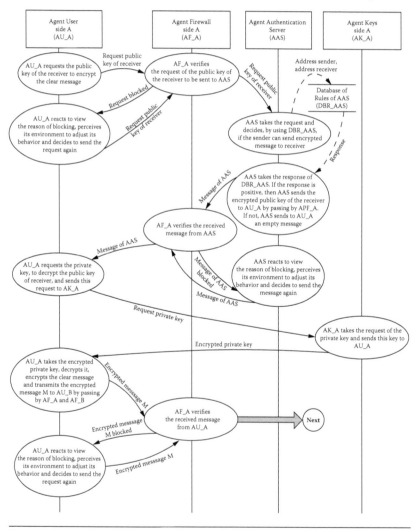

Figure 8.18(a)   Model of interactions between the different agents of the integrated MAS model concerning the transmission of encrypted data.

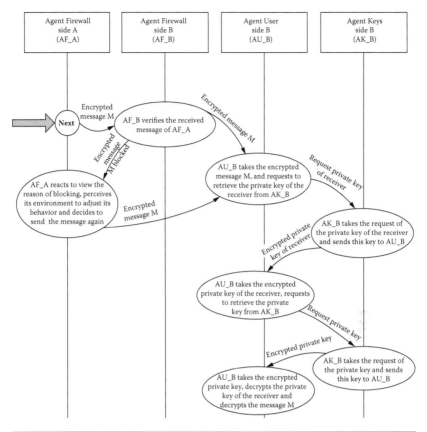

Figure 8.18(b) (continued) Model of interactions between the different agents of the integrated MAS model concerning the transmission of encrypted data.

message M. Then, it sends the message to the AU_B by passing through the AF_A and AF_B.

9. Process AR₉: the AF_A verifies the received message from the AU_A:

a. In case the message M is not valid (i.e., the type of the message sent via the IP address is not defined at the level of the outgoing port of the AF_A), the AF_A informs the AU_A that the message M is blocked. The AU_A reacts to see the cause of blocking. It perceives its environment to adjust its behavior and decides to send the message M again.

b. If not, the AF_A sends the message to the AF_B and executes the next process.

10. Process AR_{10}: the AF_B verifies the received message from the AF_A:

   a. In case the message is not valid (i.e., the type of the message sent via the IP address is not defined at the level of the incoming port of the AF_B), the AF_B informs the AF_A that the message M is blocked. The AF_A reacts to see the cause of blocking. It perceives its environment to adjust its behavior and decides to send the message M again.

   b. If not, the AF_A sends the message to the AU_B and executes the next process.

11. Process AR_{11}: the AU_B asks the AK_B to retrieve the private key of the receiver to decrypt the encrypted message M.

12. Process AR_{12}: the AK_B takes the request and sends the private key of the receiver in encrypted form to the AU_B.

13. Process AR_{13}: the AU_B takes this key and again asks the AK_B to have the private key, to decrypt the private key of the receiver.

14. Process AR_{14}: the AK_B takes the request of the private key and sends this key in encrypted form to the AU_B.

15. Process AR_{15}: the AU_B takes the encrypted private key, decrypts it, and then decrypts the encrypted message M.

8.3.5 Model of Scheduling Tasks

The MAS model described earlier represents the case of a simple system. In this system, we have a single instance of agent user side sender (AU_A) and a single instance of agent user receiver (AU_B). These two agents are involved in the exchange of encrypted information through the Internet after having the validation of system access by the biometric signature of the iris gabarit g.

Suppose we provided a complex system, in which we have multiple instances of agents AU_A and AU_B. These different instances should deal with, at the same time, the exchange of encrypted information through the Internet, after having the validation of the system access. In this case, we encounter problems of management tasks at each level of agent. To solve these problems, we propose to use the model of real-time scheduling (Birand 1991), as shown in Figure 8.19. This model

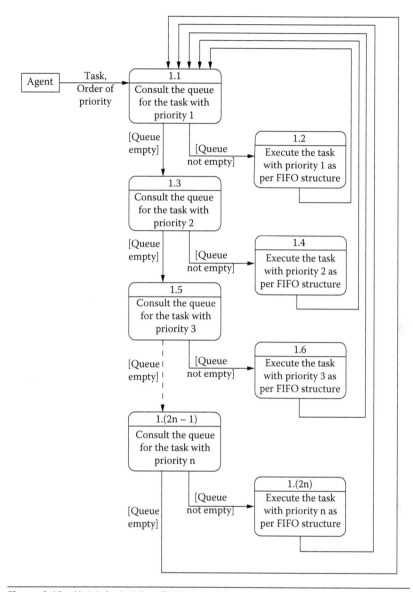

Figure 8.19 Model of scheduling of tasks in real time at the level of each agent.

consists of modeling the scheduling of tasks, at the level of each agent, according to their priority.

The sequence of the execution of these tasks according to their order of priority should be discontinued when the received message is from an agent with a higher level of importance than another. To do a permutation of tasks at the level of the concerned agent should be established.

This will lead to queue all idle tasks at the level of the agent and to carry out the urgent task after the completion of the active task.

We introduced the notion of parallelism at the level of some agents to process tasks more efficiently than in a sequential mode.

8.3.5.1 Detailed Presentation of Plan of Execution of Tasks   The model of scheduling the tasks of agents in real time is presented by a flow diagram as illustrated in Figure 8.19. This model presents the method of execution of tasks Tn (n = 1, . . . , n), for a given agent, in a consecutive way. These tasks have different priorities on a scale of 1 to n.

The sequence of all key processes Pr_i (i = 1, . . . , 4) constituting this model is described as follows:

1. Process Pr_1 (1.1): consult the processing queue of the task that has priority 1:
    a. If the queue is not empty, execute all tasks in this queue by using the structure FIFO (Padiou and Sayah 1990).
    b. If not, execute the next process.
2. Process Pr_2 (1.3): consult the processing queue of the task that has priority 2:
    a. If the queue is not empty, execute the first task in this queue that corresponds to the first entry according to the structure FIFO and then return to the process Pr_1.
    b. If not, execute the next process.
3. Process Pr_3 (1.5): in the same way, consult the processing queue that corresponds to the task with priority k (k = 3, . . . , n-1):
    a. If the queue is not empty, execute the first task in this queue that corresponds to the first entry according to the structure FIFO and then return to the process Pr_1.
    b. If not, execute the next process.
4. Process Pr_4 (1.(2n-1)): consult the processing queue of the task that has the priority n:
    a. If the queue is not empty, execute the first task in this queue that corresponds to the first entry according to the structure FIFO and then return to the process Pr_1.
    b. If not, return to the process Pr_1.

8.3.5.2 Lists of Performed Tasks by Each Agent according to Priority In this section, we include all the tasks relative to each agent of the MAS model. These tasks are established in descending order by assigning priority 1 to the one with the highest priority and so on. The priority of a task depends on its importance and the urgency of its execution. For example, if a task is closer to the goal, this task has a higher priority than others.

The decision making of the execution of some tasks in parallel is based on the concept that if two tasks are aimed at two independent objectives (e.g., a task for authentication and another for the encryption or the decryption of a message), they will be executed in parallel.

The objective of our approach to task scheduling and execution of tasks in parallel is to reduce the waiting time of tasks.

8.3.5.2.1 Tasks of Agent System User These tasks are common to both agent system users ASU_A and ASU_B to validate access to the system based on the gabarit g. These tasks T_i (i = 1, ..., 6) have different priorities in the scale of 1 to 6 as follows :

T_1: priority 1—the ASU sends to the AU the encrypted gabarit g, the percentage of the average of the gray, and the personal information.

T_2: priority 2—the ASU receives from the AU the encrypted public key. It encrypts the gabarit g and calculates the percentage of the average of gray.

T_3: priority 3—the ASU asks the AU to get the public key to encrypt the gabarit g.

T_4: priority 4—the ASU takes the parameters ri1, ri2, rp1, and rp2 to determine both external and internal edges of the iris in image Ig. It extracts the gabarit g.

T_5: priority 5—the ASU asks the AU to have the parameters ri1, ri2, rp1, and rp2 by passing it the image in grayscale Ig.

T_6: priority 6—the ASU enters the personal information. It captures the gross image containing the iris of the eye and transforms it to image in grayscale Ig.

These tasks will be performed consecutively according to their priority.

8.3.5.2.2 Tasks of Agent User   These tasks are common to both agent users AU_A and AU_B to validate access to the system based on the gabarit g. These tasks T_i (i = 1, ..., 8) have different priorities in the scale of 1 to 8 as follows:

T_1: priority 1—the AU receives from the ASU the encrypted gabarit g, the percentage of the average of gray, and the personal information and then sends to the agent verification (AV) to verify the data relative to the user in question.

T_2: priority 2—the AU sends the encrypted public key to the ASU to encrypt the gabarit g.

T_3: priority 3—the AU manages the search of the public key to encrypt the gabarit g by communicating with the agent keys (AK).

T_4: priority 4—the AU sends the parameters ri1, ri2, rp1, and rp3 to the ASU to analyze the image Ig and to get the gabarit g.

T_5: priority 5—the AU makes calculations to determine the value of the parameters ri1, ri2, rp1, and rp3.

T_6: priority 6—the AU interprets the image Ig and estimates the distance d between the user and the camera.

T_7: priority 7—the AU takes the result from AV.

T_8: priority 8—the AU records in its base of facts the performed action (identification or authentication) and the date and time of execution of this action, as well as the evaluation of result of the action (success or failure).

These tasks will be performed consecutively according to their priority.

8.3.5.2.3 Tasks of the Agent User Side A   The tasks T_i (i = 1, ..., 5) of the AU_A have different priorities in the scale of 1 to 5 as follows:

T_1: priority 1—the AU_A takes the encrypted private key and decrypts it. It encrypts the clear message to encrypted message M and sends the message M to the AF_A.

T_2: priority 2—in case the AU_A is informed by the AF_A that the message M is blocked, the AU_A reacts to see the cause of blocking. It perceives from its environment to adjust its behavior and decides to send the message M again.

T₃: priority 3—the AU_A asks the AF_A to retrieve the public key of the receiver to encrypt the clear message.

T₄: priority 4—in case the request of the public key of the receiver is blocked by the AF_A, the AU_A reacts to see the cause of blocking. It perceives from its environment to adjust its behavior and decides to send the request again.

T₅: priority 5—the AU_A takes from the AF_A the message sent by the AAS. In case the message is not empty, the AU_A asks the AK_A to retrieve the private key to decrypt the public key of the receiver.

These tasks will be performed consecutively according to their priority.

8.3.5.2.4 Tasks of Agent User Side B   The tasks T_i (i = 1, ..., 3) of the AU_B have different priorities in the scale of 1 to 3 as follows:

T₁: priority 1—the AU_B takes the encrypted private key, decrypts it, and decrypts the encrypted message M.

T₂: priority 2—the AU_B takes from the AK_B the encrypted private key of the receiver and asks for the private key to decrypt this key.

T₃: priority 3—the AU_B takes the encrypted message M and asks the AK_B to get the private key of the receiver.

These tasks will be performed consecutively according to their priority.

8.3.5.2.5 Tasks of Agent Keys   The tasks T_i (i = 1, ..., 3) of the AK have different priorities in the scale of 1 to 3 as follows:

T₁: priority 1—the AK takes the request of the private key of the receiver and sends this key in encrypted form to the AU.

T₂: priority 2—the AC takes the request of the public key to encrypt the gabarit g and sends this key in encrypted form to the AU.

T₃: priority 3—the AC takes the request of the private key and sends this key in encrypted form to the AU.

From the side B, the first two tasks, T_1 and T_2, will be performed in parallel, while the third task runs according to its priority. From the side A, the two tasks T_2 and T_3 run according to their priority.

8.3.5.2.6 Tasks of Agent Verification The tasks T_i ($i = 1, \ldots, 3$) of the AV have different priorities in the scale of 1 to 3 as follows:

T₁: priority 1—the AV manages the identification of the user in question.

T₂: priority 2—the AV manages the authentication of the user in question.

T₃: priority 3—the AV sends the result to the AU.

The first two tasks run in parallel, while the third task runs according to its priority.

8.3.5.2.7 Tasks of the Agent Authentication Server The tasks T_i ($i = 1, \ldots, 2$) of the AAS have different priorities in the scale of 1 to 2 as follows:

T₁: priority 1—the AAS takes the request and, using its base of BR_AAS, decides if this sender can send encrypted messages to the receiver:

a. In case the sender can send an encrypted message to this receiver, the AAS sends to the AF_A a message containing the encrypted public key of the receiver.

b. If not, the AAS sends to the AF_A a blank message.

T₂: priority 2—in case the AAS is informed by the AF_A that the message is blocked, the AAS reacts to see the cause of blocking. It perceives its environment to adjust its behavior and decides to send the message again.

These tasks run consecutively in the order of priority.

8.3.5.2.8 Tasks of the Agent Firewall Side A The tasks T_i ($i = 1, \ldots, 4$) of the AF_A have different priorities in the scale of 1 to 4 as follows:

T₁: priority 1—the AF_A verifies the received message from the AU_A:

a. In case the message M is not valid, the AF_A informs the AU_A that the message M is blocked.

b. If not, the AF_A sends the message to the AF_B.

T$_2$: priority 2—the AF_A verifies the received message, containing the public key or a blank message, from the AAS:

   a. In case the message is not valid, the AF_A informs the AAS that the message is blocked.

   b. If not, the AF_A sends the message to the AU_A.

T$_3$: priority 3—in case the AF_A is informed by the AF_B that the message M is blocked, the AF_A reacts to see the cause of blocking. It perceives its environment to adjust its behavior and decides to send the message M again.

T$_4$: priority 4—the AF_A verifies the request of the public key of the receiver to send it to the AAS:

   a. In case the request is not valid, the AF_A informs the AU_A that the request is blocked.

   b. If not, the AF_A sends this request to the AAS.

The two pairs of tasks, (T$_1$, T$_2$) and (T$_3$, T$_4$), run in parallel, and all of these tasks run consecutively according to their priority.

8.3.5.2.9 Tasks of Agent Firewall Side B   This agent has a single task, T$_1$, as mentioned later. In case there are several requests for this task, they are placed in the same processing queue and are executed according to the structure FIFO. The task of this agent is as follows:

T$_1$: priority 1—the AF_B verifies the received message from the AF_A containing the encrypted message M:

   a. In case the message is not valid, the AF_B informs the AF_A that the message M is blocked.

   b. If not, the AF_B sends the message to the AU_B.

8.3.6 Architecture of MAS Model

The architecture of the MAS model integrated in the IrisCryptoAgentSystem (ICAS) model, as shown in Figure 8.20, contains 11 agents. Each of these agents has its own relative properties with the tasks that should be performed. Being in a multiagent system, the agents communicate together to achieve a specific goal.

- ASU_A and ASU_B: **a**gent **s**ystems **u**sers at both ends (user U$_A$ and user U$_B$)

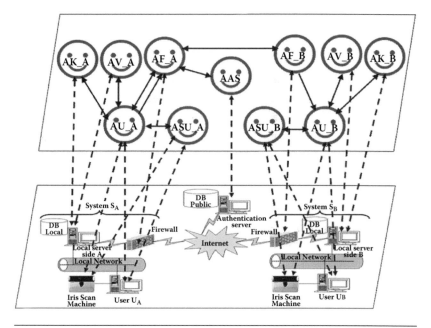

Figure 8.20   Integrated agents in our ICAS model.

- AU_A and AU_B: **a**gent **u**sers at both ends (user U_A and user U_B)
- AF_A and AF_B: **a**gent **f**irewalls at the level of the firewalls installed on both local servers on side **A** and side **B**
- AAS: **a**gent **a**uthentication **s**erver at the level of authentication server
- AK_A and AK_B: **a**gent **k**eys located at the level of the local servers on side A and side B
- AV_A and AV_B: **a**gent **v**erifications located at the level of the local servers on side A and side B

8.4 Conclusion

The wide range of expertise necessary to operate our system has naturally led us to develop a multiagent system. By introducing a modular design, it was easily possible to describe all interactions between specialized agents. On the other hand, given that our system is designed to be implemented in companies with many access points, it seems necessary for us to develop it in a MAS system.

Conclusion of Part 3

In conclusion, our model ICAS has brought efficiency in the authentication of users and the transmission of encrypted information through the Internet. Indeed, the integration of the discrete geometry has improved the processing time for the localization of external and internal edges of the iris of the eye. Our approach to the elimination of the effects of upper and lower eyelids, as well as the classification of gabarits DHVA, has made our model more pertinent. The use of intelligent systems is a rational response to the multiexpertise present in our system. In the next part, we will present the implemented parts of our ICAS model and the performed simulations.

PART 4

IMPLEMENTATION AND SIMULATIONS

The methods that we propose to enhance the algorithm of iris recognition are presented in this part of the book. The immediate application of these methods is described by algorithms in pseudocode form. The simulations illustrate the operation of these algorithms. A proposal for a study of implementation of our model IrisCryptoAgentSystem (ICAS) in hospital services is detailed.

9

IMPLEMENTATION OF NEW METHODS

The proposed model described in this chapter is based on the biometric authentication method using the iris of the eye and the asymmetric cryptography using the Rivest–Shamir–Adleman (RSA) algorithm (Stallings 1999). This model is founded on the use of the biometric iris signature gabarit of a person to access a computer system. The justification for the choice of the biometric model based on the iris of the eye is presented in this chapter, as well as the description of the different modules constituting this model. In this model the authentication process allows obtaining a gabarit "diagonal horizontal vertical approximation" (DHVA)* encrypted. This will be compared with other gabarits DHVA encrypted stored in the database to verify access to the computer system. A method of classification of these gabarits is detailed in order to speed the verification process, especially for a large database.

9.1 Presentation of Our Algorithm for Iris Recognition

In this section, we present the parts of the algorithm for iris recognition of our conception.

9.1.1 Algorithm of the Optimized Circular Hough Transform

The algorithm of the optimized circular Hough transform is represented by the pseudocode illustrated in Figure 9.1. This algorithm consists of searching a circle (i.e., external edge or internal edge contour of the iris) of radius R in the binary image (cf. Chapter 7, Section

---

* This name is inspired from the concept of approximation of the Haar wavelet.

```
Function HoughOptimized(BImage,R)

NbLin  ⟵  NbofLin(BImage[i,j])

NbCol  ⟵  NbofCol(BImage[i,j])

Accum[i,j]  ⟵  0

LCoords[i']  ⟵  FindLC(BImage[i,j])

CCoords[j']  ⟵  FindCC(BImage[i,j])

n ⟵ length(CCoords[j'])

For counter = 1 to n step = 1 do
    ColCoord  ⟵  CCoords[counter]
    j₁ ⟵ ColCoord − R
    j₂ ⟵ ColCoord + R

    if j₁ < 1 then
        j₁ ⟵ 1
    endif

    if j₂ > NbCol then
        j₂ ⟵ NbCol
    endif

    for j₀ = j₁ to j₂ step = 1 do
        temp  ⟵  ((R)²-(NbCol-j₀)²)
        i₀₁ ⟵ round(LCoords[counter] − sqrt(temp))
        i₀₂ ⟵ round(LCoords[counter] + sqrt(temp))

        if i₀₁ < NbLin & i₀₁ > = 1 then
            Accum[i₀₁,j₀] ⟵ Accum[i₀₁,j₀]+1
        endif

        if i₀₂ < NbLin & i₀₂ > = 1 then
            Accum[i₀₂,j₀] ⟵ Accum[i₀₂,j₀]+1
        endif
    endfor
endfor
```

Figure 9.1   Pseudocode of the function of the optimized circular Hough transform.

7.4.3.1) BImage. This image is generated by the operator of Canny and represented as a matrix BImage[i,j].

To reduce the search time, we considered only the elements of the matrix BImage[i,j] with a value of −1. For these elements, we kept, respectively, the coordinate lines and the coordinate columns in two vectors LCoords[i'] and CCords[j']. The values of these two vectors are determined by the two functions FindLC (BImage[i,j]) and FindCC (BImage[i,j]).

To seek the circle corresponding to the final edge (i.e., external edge or internal edge of the iris), we applied the equation of a circle defined by

$$x^2 + y^2 = R^2 \tag{9.1}$$

where

x is represented by two symmetrical line coordinates i_{01} and i_{02} in the matrix BImage[i,j].

y is represented by the column coordinate j_0 in the matrix BImage[i,j] and the value of j_0 belongs to the range [1 ... NbCol], where NbCol represents the number of columns of the matrix BImage[i,j]. This number of columns is determined by the function NbofCol(BImage[i,j]).

We calculated for all elements of coordinates j_0 the symmetrical line coordinates i_{01} and i_{02}.

We incremented by one the value of the element Accum[i_{01},j_0] or Accum[i_{02},j_0] for all values of i_{01} and of i_{02}, which are strictly positive and less than the number of rows NbLin of the matrix BImage[i,j]. The latter is determined by the function NbofLin(BImage[i,j]).

The circle, corresponding to the edge in question, is characterized by the radius R and passing through the point relative to the element with the first maximum value in the matrix Accum[i,j].

9.1.2 Algorithm of the Determination of the Raster T

The algorithm for determining the raster T is represented by the pseudocode illustrated in Figure 9.2. This algorithm consists of searching the raster T that corresponds to the region of intersection

```
Function DeterminationRasterT (ρ_iris, ρ_pupil, Is)

a  ⟵  ρ_iris + 10
b  ⟵  ρ_iris − 10
θ'  ⟵  arctan ((b/a)* sqrt((a² − ρ²_iris)/(ρ²_iris − b²)))
ρ_cc  ⟵  (ρ_iris − ρ_pupil + 1)
Iris[i,j] ⟵ Is
T[i,j] ⟵ −1

for θ = 0 to θ' step = 0.1 do
   for ρ = 1 to ρ_cc step = 1 do
      T[i,j] ⟵ elements Iris[i,j] correspondent to the section of circular donut from 0 to θ'
   endfor
endfor

for θ = (θ' + 0.1) to (π − θ') step = 0.1 do
   ρ_ce ⟵ ((a*b)/sqrt((b* cosθ)² + (a*sinθ)²))
   for ρ = 1 to (ρ_ce − ρ_pupil + 1) step = 1 do
      T[i,j] ⟵ elements Iris[i,j] correspondent to the section of circular donut from ( θ' + 0.1) to (π − θ')
   endfor
endfor

for θ = (π − θ' + 0.1) to (π + θ') step = 0.1 do
   for ρ = 1 to ρ_cc step = 1 do
      T[i,j] ⟵ elements Iris[i,j] correspondent to the section of circular donut from (π − θ' + 0.1) to (π + θ')
   endfor
endfor

for θ = (π + θ' + 0.1) to (2π − θ') step = 0.1 do
   ρ_ce ⟵ ((a*b)/sqrt((b* cosθ)² + (a*sinθ)²))
   for ρ = 1 to (ρ_ce − ρ_pupil + 1) step = 1 do
      T[i,j] ⟵ elements Iris[i,j] correspondent to the section of circular donut from (π + θ' + 0.1) to (2π − θ')
   endfor
endfor

for θ = (2π − θ' + 0.1) to 2π step = 0.1 do
   for ρ = 1 to ρ_cc step = 1 do
      T[i,j] ⟵ elements Iris[i,j] correspondent to the section of circular donut from (2π − θ' + 0.1) to 2π
   endfor
endfor

Return T[i,j]
```

Figure 9.2   Pseudocode of the function of determination of the raster T.

between the circular ring Is of the iris and an ellipse to eliminate the effects of the upper and lower eyelids.

The circular ring is characterized by the radii ρ_{iris} of the external edge and ρ_{pupil} of the internal edge, and the distance ρ_{cc} between these two edges. It is represented in polar coordinates in the form of a matrix Iris[i,j], with

- i: number of rows of the matrix
- j: number of columns of the matrix

The ellipse is characterized by the larger radius a (equal to ρ_{iris} + 10) and the small radius b (equal to ρ_{iris} − 10) and the variable polar radius ρ_{ce}.

We calculated the fixed angle θ' (cf. Chapter 7, Section 7.4.3.2), which determines the four points of intersection between the external edge of the ring and the ellipse. The intersection region is divided into two sections of the following:

- Left and right circular ring characterized by a radius ρ belonging to the interval $[1 \ldots \rho_{cc}]$
- Upper and lower elliptical ring characterized by a radius ρ belonging to the interval $[1 \ldots (\rho_{ce} - \rho_{pupil+1})]$

The raster T is represented as a matrix $T[i,j]$ having the same size as the matrix $Iris[i,j]$. This raster is initialized to -1 to hide the part that does not correspond to the region of intersection between Is and the ellipse.

We assigned to that raster the elements $Iris[i,j]$ relative to the intersection region as follows:

- The elements of $Iris[i,j]$ corresponding to the circular ring section of 0 to θ'
- The elements of $Iris[i,j]$ corresponding to the elliptical ring section of $(\theta' + 0.1)$ to $(\pi - \theta')$
- The elements of $Iris[i,j]$ corresponding to the circular ring section of $(\pi - \theta' + 0.1)$ to $(\pi + \theta')$
- The elements of $Iris[i,j]$ corresponding to the elliptical ring section of $(\pi + \theta' + 0.1)$ to $(2\pi - \theta')$
- The elements of $Iris[i,j]$ corresponding to the circular ring section of $(2\pi - \theta' + 0.1)$ to 2π

This function returns the value of the matrix $T[i,j]$, which represents the raster T. From this raster we extracted, by applying the two-dimensional discrete Haar transform, the gabarit DHVA.

9.1.3 Algorithm for the Classification of Gabarits DHVA

The algorithm for the classification of the gabarits DHVA is divided into two parts: the registration of a gabarit and the search for a gabarit.

9.1.3.1 Algorithm to Record a Gabarit DHVA   The algorithm for recording a gabarit DHVA, designated by g, is represented by the pseudocode illustrated in Figure 9.3. This algorithm is to record a given template g_d in a hashing file. This gabarit is indexed by the index of the percentage of the average of gray $\overline{Mgucp}$. This index is an element of the leaf (or node) $Bcp[i]$ $(i = 1 \ldots N)$ pointed by a root index $Root[p]$ $(p = 1 \ldots \infty)$, which exists at the level of the root of the tree,

Procedure RecordingGabarit(g_d, $\overline{M}$gucp, Root[p], Bcp[i], N)

Bcp[i] ⟵ insert a new index $\overline{M}$gucp related to g_d

Recording the gabarit g_d, indexed by $\overline{M}$gucp, in the hashing file containing the gabarits g

N' ⟵ length(Bcp[i])

if N' > N **then**

  M ⟵ find the position of the middle of the node Bcp[i]

  Root[p] ⟵ add Bcp[M]

  Bcp_1[N] ⟵ 0

  j ⟵ 1

  for h = 1 **to** (M−1) **step** = 1 **do**

   Bcp_1[j] ⟵ Bcp[h]

   j ⟵ j+1

  endfor

  Bcp_2[N] ⟵ 0

  j ⟵ 1

  for h = M **to** N' **do**

   Bcp_2[j] ⟵ Bcp[h]

   j ⟵ j+1

  endfor

endif

Figure 9.3   Pseudocode of the procedure of recording of the gabarit DHVA.

where N represents the maximum index number at the level of the leaf of the tree.

After insertion of $\overline{M}$gucp relative to g_d, if the number of elements N' of the leaf Bcp[i] exceeds N, we apply the method of dividing the node into two nodes, Bcp_1[i] and Bcp_2[i]. We seek the middle index Bcp[M] corresponding to the position M in the node Bcp[i]. We add in Bcp_1[i] the elements Bcp[h], where h = 1 ... (M-1), and in Bcp_2[i] the elements Bcp[h], where h = M ... N'.

9.1.3.2 Algorithm to Search for a Gabarit DHVA   The search algorithm of a gabarit DHVA is represented by the pseudocode illustrated in Figure 9.4. This algorithm consists of fetching in the root of the tree Root[p] if the index t, equal to (int($\overline{M}g_d$cp) + 1), exists in the root.

Procedure FindGabarit(g_d, $\overline{Mg_1}cp$, Root[p], Bcp[i])

Bool ⟵ find in Root[p] the indext = (integer($\overline{Mg_1}cp + 1$)

If Bool = 0 **then**

  Bool1 ⟵ find in Bcp[i] indexed by (t−1) if the rate of dissimilarity between $\overline{Mg_1}cp$ and those in BCP[i] is < 0.0200

  If Bool1 = 0 **then**

    Bool2 ⟵ find in Bcp[i] indexed by (t+1) if the rate of dissimilarity between $\overline{Mg_1}cp$ and those in BCP[i] is < 0.0200

    If Bool2 = 0 **then**

      Bool3 ⟵ find in Bcp[i] indexed by (t−2) if the rate of dissimilarity between $\overline{Mg_1}cp$ and those in BCP[i] is < 0.0200

      If Bool3 = 0 **then**

        Bool4 ⟵ find in Bcp[i] indexed by (t+2) if the rate of dissimilarity between $\overline{Mg_1}cp$ and those in BCP[i] is < 0.0200

        If Bool4 = 0 **then**
          display "Person rejected" in case of authentication
            or "New person" in case of identification
        else
          display "Person accepted" in case of authentication
            or "Person already recorded" in case of identification
        endif

      else
        display "Person accepted" in case of authentication
          or "Person already recorded" in case of identification
      endif

    Sinon
      display "Person accepted" in case of authentication
        or "Person already recorded" in case of identification
    Finsi

  Sinon
    display "Person accepted" in case of authentication
      or "Person already recorded" in case of identification
  endif

else

  Bool5 ⟵ findinBcp[i] indexed by t if the rate of dissimilarity between $\overline{Mg_1}cp$ and those in BCP[i] is < 0.0200

  If Bool5 = 0 **then**
    display "Person rejected" in case of authentication
      or "New person" in case of identification
  endif
    display "Person accepted" in case of authentication
      or "Person already recorded" in case of identification
  endif

endif

Figure 9.4 Pseudocode of the search procedure of the gabarit DHVA.

If the index t does not exist, we seek, respectively, in the base of neighborhood of t constituted of the indexes (t − 1), (t + 1), (t − 2), and (t + 2). Otherwise, we look in the node Bcp[i] indexed by t if the rate of dissimilarity between the given g_d having a percentage of average of gray $\overline{Mg_d}cp$ and the gabarits stored in the node is less than 0.0200.

If this rate is less than 0.0200, the two compared gabarits are identical. Otherwise, these two gabarits are considered to be different.

9.2 Analysis of Performance

The performance analysis is based on the concept of the theory of calculation of complexity (Xuong 1992). We applied this concept to calculate the time required for the performance of the proposed algorithm after modifying the circular Hough transform, as shown in Figure 9.5.

We calculated the time of complexity $T(n)$ of the algorithm by applying the rules of simplification (Xuong 1992) as follows:

$$T(n) = O(1) + O(1) + O(1) + O(1) + O(1) + O(1) + O(n) *$$
$$\{[O(1) + O(1) + O(1)] + [O(1) * O(1)] + [O(1) * O(1)] + O(j2) *$$
$$[[O(1) + O(1) + O(1)] + [O(1) * O(1)] + [O(1) * O(1)] + [O(1) *$$
$$O(1)]]\} = O(n^*j_2) \qquad (9.2)$$

Our algorithm is of the order n^*j_2, designated by $O(n^*j_2)$, where

- n: maximum coordinate of the column in the matrix BImage[i,j] for the elements which have a value equal to one
- j_2: upper bound of the interval $[j_1,j_2]$ relative to the column coordinates of the matrix BImage[i,j] to find a circle (i.e., possibility to locate an edge)

The other algorithms are of the order j^*j_2, designated by $O(j^*j_2)$, where

- j: maximum coordinate of the column in the matrix BImage[i,j]
- Top of form

The comparison of the time of complexity of our algorithm $T(n)$ with the time of complexity of other algorithms $T_1(n)$ is based on the concept of asymptotic estimation (Xuong 1992).

In our algorithm $n \leq j$ and $j_2 \leq n$, the time complexity of our algorithm is of the order $O(n^2)$ and that of the other algorithms is of the order $O(j^2)$.

Function HoughOptimized(BImage, R)	Order
	$\underbrace{}$
NbLin ⟵ NbofLin(BImage[i,j])	//O(1)
NbCol ⟵ NbofCol(BImage[i,j])	//O(1)
Accum[i,j] ⟵ 0	//O(1)
LCoords[i'] ⟵ **FindLC(BImage[i,j])**	//O(1)
CCoords[j'] ⟵ **FindCC(BImage[i,j])**	//O(1)
n ⟵ length(CCoords[j'])	//O(1)
for counter = 1 **to** n **step** = 1 **do** //n ≤ j	//O(n)
ColCoord ⟵ CCoords[counter]	//O(1)
j_1 ⟵ ColCoord − R	//O(1)
j_2 ⟵ ColCoord + R	//O(1)
if j_1 < 1 **then**	//O(1)
j_1 ⟵ 1	//O(1)
endif	
if j_2 > NbCol **then**	//O(1)
j_2 ⟵ NbCol	//O(1)
endif	
for j_0 = j_1 **to** j_2 **step** = 1 **do** //j_2 ≤ n	//O(j_2)
temp ⟵ $((R)^2-(NbCol-j_0)^2)$	//O(1)
i_{01} ⟵ round(LCoords[counter] − sqrt(temp))	//O(1)
i_{02} ⟵ round(LCoords[counter] + sqrt(temp))	//O(1)
if i_{01} < NbLin & i_{01} > = 1 **then**	//O(1)
Accum[i_{01},j_0] ⟵ Accum[i_{01},j_0]+1	//O(1)
endif	
if i_{02} < NbLin & i_{02} > = 1 **then**	//O(1)
Accum[i_{02},j_0] ⟵ Accum[i_{02},j_0]+1	//O(1)
endif	
endfor	
endfor	

Figure 9.5 Pseudocode of the function of the optimized circular Hough with the time of each instruction.

In the worst case, $j = n$, and both algorithms have the same complexity $O(j^2)$. At best, $n < j$.

In practice,

$$n \leq \frac{j}{3}$$

This implies a time of complexity of our algorithm of the order

$$O\left(\frac{j^2}{9}\right)$$

In other words, the computation time of our algorithm is reduced to 10% by comparing with other algorithms. This estimation of the computation time is based on different simulations detailed in the next chapter. We can conclude that our algorithm is more efficient than others.

9.3 Conclusion

In summary, we have presented in this chapter the algorithms of the methods that we propose for improving our algorithm of iris recognition. The theoretical performance analysis showed in practice the efficiency of our algorithm in terms of processing time compared to other algorithms.

10

SIMULATION OF MODULES

This chapter aims to present different simulations proposed for the localization of the external and internal edges of the iris of the eye, to eliminate the effects of the upper and lower eyelids that can hide an important part of the iris, as well as the classification of the gabarits diagonal horizontal vertical approximation (DHVA) (cf. Chapter 7, Section 7.4.3.3). We make a comparative analysis and evaluation of the results of these simulations.

10.1 Simulations and Analysis of the Edges Localization Module

In this section we present a simulation of our conceived module for the localization of external and internal edges of the iris. This module showed time efficiency (cf. Chapter 5, Section 5.3.2) with a reduction of processing time to 5.5 seconds on average in comparison with the fifth simulation (cf. Chapter 5, Section 5.2.5).

Our tests were conducted on a sample of 257 images taken from three sources of images:

1. Random images from Internet documents
2. Images taken by the Sony CCD color camera with a resolution of 520 TV lines
3. Images from the CASIA iris image database (CASIA 2006)

Three different types of images are respectively illustrated by (a), (b), and (c) in Figure 10.1.

We used MATLAB R2006a on a Pentium IV with a processor of 2.2 GHz and 1 MB of RAM to perform these simulations.

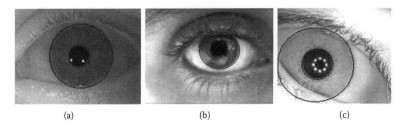

(a)   (b)   (c)

Figure 10.1   Localization of external and internal edges by the circular Hough transform.

10.1.1 Simulation of Our Module

This simulation is based on the application of the circular Hough transform for the localization of external and internal edges of the iris shown in the fifth simulation detailed in Chapter 5 (cf. Section 5.2.5).

The fifth simulation showed good accuracy in the localization of external and internal edges (Figure 10.1) for several states of the iris (i.e., rotation, illumination level) and a processing time of 13.5 seconds on average. Our module focuses on time efficiency (cf. Chapter 5, Section 5.3.2). We look in the binary image only for the elements that have a value equal to one (cf. Chapter 7, Section 7.4.3.1). This reduces the time taken to locate edges to 5.5 seconds. Hence, the duration of treatment is 8 seconds on average.

10.1.2 Analysis of the Simulations Based on Time Aspects

Our study focuses on the iris of the eye, which is represented geometrically by a ring bounded by two conical (i.e., circle) C1 and C2 (cf. Chapter 5, Section 5.3.2). For a given image, we focus on external or internal edges of the iris, and their localization (cf. Chapter 2, Section 2.6).

Consider a finite family of images F composed of a product of three families (F_1, F_2, F_3):

$$F = F_1 \times F_2 \times F_3 \tag{10.1}$$

- F_1 is a set of 35 images taken randomly from Internet documents.
- F_2 is a set of 72 images taken by the Sony CCD camera with a resolution of 520 TV lines.
- F_3 is a set of 150 images issued from the CASIA iris image database (CASIA 2006).

For each image belonging to the families Fj, we should determine the external edge C1 and the internal edge C2 previously defined to determine the processing time of the localization process.

Both edges are considered as a circle in the methods used in the two simulations: simulation 1 and simulation 2, where

- Simulation 1 represents the fifth simulation (cf. Chapter 5, Section 5.2.5), and
- Simulation 2 represents the simulation of our proposed model (cf. Chapter 7, Section 7.4.3.1).

Let:

- n_1 represent the cardinality of F_1, n_1 = card (F_1)
- n_2 represent the cardinality of F_2, n_2 = card (F_2)
- n_3 represent the cardinality of F_3, n_3 = card (F_3)

The time efficiency consists of comparing the effectiveness of a method M_k (k = 1, 2) versus another of the three families F_1, F_2, and F_3. This efficiency will be calculated later.

We pose $t_{h,k}^{j}$ the computation time to determine the edges C1 and C2 of the iris number h belonging to the family Fj (j = 1, ..., 3) and for the method M_k, as shown in Table 10.1. This gives us an estimation of average time:

$$\overline{T}_{j,k} = \frac{1}{n_j} \sum_{h=1}^{n_j} t_{h,k}^{j} \qquad (10.2)$$

and we have

$$\overline{T}_j = \sum_{k=1}^{5} \overline{T}_{j,k} \qquad (10.3)$$

Table 10.1 Time Values of the Localization Process of External and Internal Edges $t_{h,k}^{j}$ for Each Method and by Family Fj

	F_1	F_2	F_3
M_1	5	18.5	17
M_2	4	10.5	9.5

The calculation of the time efficiency of the method M_k on the family Fj for determining the edges C1 and C2 is

$$T_{j,k}^e = 1 - \frac{\overline{T}_{j,k}}{\overline{T}_j} \qquad (10.4)$$

We get Table 10.2 of time efficiency from our sample on the families F_1, F_2, and F_3, with

- Card (F_1) = 35
- Card (F_2) = 72
- Card (F_3) = 150

The corresponding histogram is seen in Figure 10.2.

We observe that the higher the time efficiency rate is, the more efficient the method is from the point of view of computation time for the localization of the edges C1 and C2.

In view of the results obtained at the level of time efficiency, the method M_2 does not have a big difference compared to the method M_1 ($T_{j,k}^e = 0.53$) for the localization of the edges C1 and C2 on the family F_1. However, the method M_2 is clearly the most efficient compared to the method M_1 on both families F_2 and F_3.

Table 10.2 Values $T_{j,k}^e$

	F_1	F_2	F_3
M_1	0.47	0.36	0.36
M_2	0.53	0.64	0.64

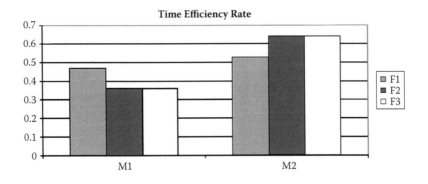

Figure 10.2 Time efficiency rate relative to each method on the three families.

We can deduce that our method M_2 for the localization of the external and internal edges of the iris of the eye is the most pertinent and effective (i.e., temporal aspect) compared to other methods used (cf. Chapter 5, Section 5.2).

10.2 Simulations and Analysis of the Elimination of Eyelids' Effects Module

In this section, we will present three simulations based on different concepts proposed for the elimination of the effects of the upper and lower eyelids, in order to choose the most appropriate method. Our tests are performed on a sample of 257 images taken from the CASIA iris database image (CASIA 2006). The three following simulations are performed on four samples of images:

a. The left iris of person 1
b. The left iris of person 1 with one position different from the preceding
c. The right iris of person 1
d. The right iris of person 2

The binary gabarits (cf. Chapter 2, Section 2.5.4) due to Daugman and the gabarits of our model "gabarit DHVA" issued from these different types of images will be represented in the three following simulations by gabarit(a), gabarit(b), gabarit(c), and gabarit(d).

The results of the comparison between two binary gabarits are obtained by using the Hamming distance (HD), while the results of the comparison between two gabarits of our model are obtained by using the method of comparing two vectors (cf. Chapter 7, Section 7.4.5.2). These results present the rate of dissimilarity (cf. Chapter 6, Section 6.2) between two given gabarits. Recall that we also used the discrete Haar transform in two dimensions and one level for the extraction of iris biometric features gabarit (cf. Chapter 2, Section 2.5.4).

10.2.1 First Simulation: Two Sections of Circular Rings to Left and to Right of 150°

This simulation consists of examining two sections of a circular ring, including the iris, centered at the center of the pupil and arranged

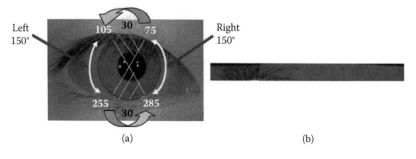

(a)   (b)

Figure 10.3   Two sections of circular rings to left and to right of 150°.

symmetrically on the left and right of the region of the iris. These two
sections each includes an angle of 150° (i.e., the two parts not in the
hashed region of the iris), as shown in Figure 10.3.

*10.2.1.1 Results Obtained on Rate of Dissimilarity between Binary Gabarits
a, b, c, and d*

Rate(gabarit(a),gabarit(b)) = 0.0219
Rate(gabarit(a),gabarit(c)) = 0.0191
Rate(gabarit(a),gabarit(d)) = 0.0216
Rate(gabarit(c),gabarit (d)) = 0.0218

*10.2.1.2 Results Obtained on Rate of Dissimilarity between Gabarits DHVA
a, b, c, and d*

Rate(gabarit(a),gabarit(b)) = 0.0641
Rate(gabarit(a),gabarit(c)) = 0.0606
Rate(gabarit(a),gabarit(d)) = 0.0633
Rate(gabarit(c),gabarit(d)) = 0.0635

10.2.2 Second Simulation: One Circular Ring Section of 330°

This simulation carries out the analysis of one section of a circular
ring of 330° contained in the iris, centered at the center of the pupil
and located in the lower region of the iris (i.e., not part of the hashed
region of the iris), as shown in Figure 10.4.

Part kept 330°

(a)   (b)

Figure 10.4 One section of circular ring of 330°.

10.2.2.1 Results Obtained on Rate of Dissimilarity between Binary Gabarits a, b, c, and d

Rate(gabarit(a),gabarit(b)) = 0.0183
Rate(gabarit(a),gabarit(c)) = 0.0159
Rate(gabarit(a),gabarit(d)) = 0.0183
Rate(gabarit(c),gabarit(d)) = 0.0184

10.2.2.2 Results Obtained on Rate of Dissimilarity between Gabarits DHVA a, b, c, and d

Rate(gabarit(a),gabarit(b)) = **0.0557**
Rate(gabarit(a),gabarit(c)) = 0.0521
Rate(gabarit(a),gabarit(d)) = **0.0546**
Rate(gabarit(c),gabarit(d)) = 0.0551

10.2.3 Third Simulation: Intersection between Region
of the Iris and a Given Ellipse

This simulation consists of taking a semielliptical ring that is a part of the iris (i.e., not part of the hashed area of the iris), as shown in Figure 10.5. This area includes the inner part of a given ellipse centered at the center of the pupil and that cuts the iris. According to the simulations performed on these 257 images, we have chosen the values for large radius of the ellipse designated by a = radius iris + 10 and the smallest radius of the ellipse designated by b = radius iris − 10.

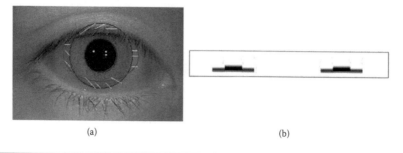

(a)   (b)

Figure 10.5   Intersection part between the region of the iris and a given ellipse.

10.2.3.1 Results Obtained on Rate of Dissimilarity between Binary Gabarits a, b, c, and d

Rate(gabarit(a),gabarit(b)) = 0.0139
Rate(gabarit(a),gabarit(c)) = 0.0160
Rate(gabarit(a),gabarit(d)) = 0.0151
Rate(gabarit(c),gabarit(d)) = 0.0175

10.2.3.2 Results Obtained on Rate of Dissimilarity between Gabarits DHVA a, b, c, and d

Rate(gabarit(a),gabarit(b)) = 0.0199
Rate(gabarit(a),gabarit(c)) = 0.0220
Rate(gabarit(a),gabarit(d)) = 0.0207
Rate(gabarit(c),gabarit(d)) = 0.0227

10.2.4 Evaluations of Simulations

The simulations we conducted show that the rate of dissimilarity is very similar, as illustrated in Table 10.3. We find that the rates of dissimilarity are very low in binary mode, which gives this mode mediocre results. For binary mode, on the second simulation, we see that the value of the rate of dissimilarity for two identical or different irises is the same: 0.0183.

Regarding the first and the second simulation, the rate of dissimilarity for two identical irises is greater than that for two different irises. The third simulation, based on the intersection between the iris region and a given ellipse of our model, gives consistent and very satisfactory results. This method provides a better comparison between

Table 10.3   Results of Rate of Dissimilarity on Different Simulations

RATE OF DISSIMILARITY	TYPE OF GABARIT	GABARITS SAME PERSON DIFFERENT POSITIONS	GABARIT LEFT AND GABARIT RIGHT SAME PERSON	GABARIT LEFT AND GABARIT RIGHT DIFFERENT PEOPLE	GABARIT RIGHT AND GABARIT RIGHT DIFFERENT PEOPLE
First simulation	Binary	0.0129	0.0191	0.0216	0.0218
	Our model	0.0641	0.0606	0.0633	0.0635
Second simulation	Binary	**0.0183**	0.0159	**0.0183**	0.0184
	Our model	**0.0557**	0.0521	**0.0546**	0.0551
Third simulation	Binary	0.0139	0.0160	0.0151	0.0175
	Our model	0.0199	0.0220	0.0207	0.0227

irises. Experimentally, two irises are identified as identical when the rate of dissimilarity is less than or equal to 0.0200.

10.3 Simulations and Analysis of Classification of Gabarits DHVA Module

In this section, we present a simulation of our module conceived for the classification of gabarits DHVA. This module is based on the indexed hierarchy in order to speed up the access time on a large database. Our tests are performed on a sample of 257 different iris images from the CASIA iris image database (CASIA 2006).

We calculated the percentage of the average of gray of all required gabarits DHVA. We used the application Oracle 9i Developer for the development of the grouping algorithm and the search algorithm for a given gabarit DHVA, and a management system database Oracle 9i Database to record data. These simulations gave the following results:

- The set of couples ($\overline{M}$gcp, nbe), in which $\overline{M}$gcp represents the value of the key at the level of the root and nbe the number of elements obtained in each group p (cf. Chapter 7, Section 7.4.5.1):

  {(14,1), (15,1), (16,1), (17,2), (18,4), (19,4), (20,1), (21,8), (22,7), (23,9), (24,10), (25,13), (26,23), (27,20), (28,12), (29,31), (30,25), (31,20), (32,16), (33,10), (34,7), (35,7), (36,8), (37,1), (38,2), (39,3), (40,1), (45,1)}

- The values of $\overline{\text{Mgucp}}$ represent the indexes at the level of the indexed leaf by the index $\overline{\text{Mgcp}}$ of the key for a group p at the level of the root r (cf. Chapter 7, Section 7.4.5.1). These values are between 13.29 and 44.62.
- The values of $\overline{\text{Mgucp}}$ for the same iris, but with three different illumination levels (less clear: trend toward black [equal to 0], mean, more clear: trend toward white [equal to 255]). These values are represented, for example, as follows:
  - gabarit 1 has the triplet (26.64, 27.00, 28.50).
  - gabarit 2 has the triplet (26.42, 27.94, 28.11).

In view of these results, we decided that the optimal value of N (cf. Chapter 7, Section 7.4.5.1), representing the maximum number of indexes at the level of the leaf, is equal to 50. We also justify our choice for the base of neighborhoods (cf. Chapter 7, Section 7.4.5.1) of a given element gabarit DHVA.

10.4 Conclusion

In this chapter we have detailed methods for the localization of external and internal edges of the iris, to eliminate the effects of upper and lower eyelids, as well as the classification of gabarits DHVA. The obtained results justify the performance of the methods presented in our biometric model (cf. Chapter 7, Sections 7.4.3.1 and 7.4.3.2).

11

APPLICATION

This chapter aims to present a practical application of our model IrisCryptoAgentSystem (ICAS), which will be implemented in hospital services. We show a scenario describing the information system and the constraints with the privileges and access rights of users of the system. We make a study of resources for the proper functioning of this application.

11.1 Description of the Information System

- The information system meets the needs of a group of doctors who wish to access their own files and their patients' records.
- Each doctor has his or her own office and works in the clinic.
- The access of the doctors to the relevant files is done locally or remotely.
- The local access is done in the office or in the clinic.
- The remote access is in both directions between the clinic and the office through the Internet.
- The clinic staff can access all records of doctors regarding their patients' hospitalizations. This aims to print a report summarizing a given patient's hospitalization or intervention that the patient has undergone.
- The chiefs of departments of the clinic have the right to access the records of all patients in their service.
- The patients do not have the ability to access the system for confidentiality reasons.
- If a doctor needs information about an intervention performed by another doctor, he or she does not have the right to inspect the files of intervention. The doctor asks directly for the information from the other doctor.

- Similarly, a chief of department has no access to patient records from other services. Such information is requested from other chiefs of departments.
- Each user should authenticate using his or her iris and his or her personal information to gain access to confidential information.

The required personal data for this user to be able to access the system are as follows:

- The user's name
- The name of the user's father
- The name of the user's mother
- The user's birth date
- The user's gender
- The user's address
- The user's phone number
- The user's e-mail address

11.2 Schema of Granted Privileges by Users

Figure 11.1 shows the assigned privileges for the local or remote access users to the system. The privilege "select" consists of read-only information for reasons of display or printing. This privilege

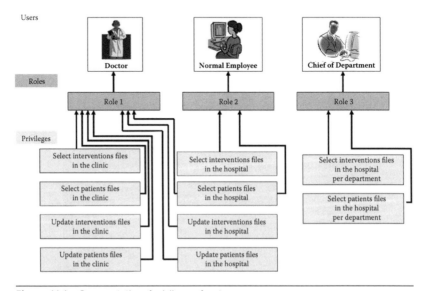

Figure 11.1   Representation of privileges of system users.

should be limited for employees and conditional for the chiefs of departments in the clinic. The privilege "update" is to add, modify, or delete information.

The three roles (role 1, role 2, and role 3) include each set of privileges. Each privilege can be assigned to multiple roles. Each role is assigned to a category of users (doctor, normal employee, or chief of department).

The doctors have more privileges than other users accessing the system.

11.3 Study of the Resources

For the realization and implementation of our proposed model for the clinical services, we need a set of human resources, hardware, and software.

11.3.1 Human Resources

We need a team of three engineers. Two engineers are necessary for the implementation of the complete system over a period of 1 year. These engineers must be qualified in image processing, databases, and object-oriented programming. An engineer for maintenance should have experience in qualification of services and telecommunications.

11.3.2 Software

We propose two software options for the implementation of the system.

11.3.2.1 Option 1
- Java for the web applications
- PHP
- MySQL for the management of the database

This option could be installed free from the Internet as open sources.

11.3.2.2 Option 2
- MATLAB R2006 and higher version
- Oracle database 9i and higher version of Oracle database to manage the database

This option is costly according to the number of licenses per user.

The second option is more advantageous than the first from the point of view of

- Data security due to the application Oracle database
- Fast handling due to some automatic functions generated by the software MATLAB for image processing

11.3.3 Hardware

Each workstation should be equipped with a color or black and white camera. The camera should have a good resolution to provide a suitable image for a good localization of edges of the iris. The operating system should be Windows XP Professional, Windows Vista, or Windows 7 Professional. The minimum specifications of the machine are a Pentium IV processor with 2.2 GHz, 1 MB of memory, and an 8 GB hard drive expandable depending on the size of the data. However, the market offers machines with high specifications that provide high performance related to the processing time.

11.4 Conclusion

In this chapter we presented the scientific aspects for the application of our model ICAS in the hospital services. This should enable the realization of our model in hospital areas.

Conclusion of Part 4

In summary, the methods we propose for enhancing of our algorithm of iris recognition have shown their effectiveness (e.g., processing time) compared to other algorithms. The results of simulations justify these methods by their performance in our model ICAS. A proposal for a study for the implementation of our model ICAS in hospital services allows for validating its realization.

Conclusion and Perspectives

The study from the extensive bibliography on the field of secure access to confidential data shows that our model, IrisCryptoAgentSystem (ICAS), is topical and brings new elements. This model is based on the biometric technique using the iris of the eye for the authentication of users and the asymmetric cryptography method using the Rivest–Shamir–Adleman algorithm to encrypt the information. This model ensures secure access to confidential data (i.e., stored in the database or transmitted through the network).

The model ICAS is developed in a multiagent system (MAS) consisting of agents of different types (e.g., biometric, cryptographic, etc.). These agents interact in a coherent way to manage the operation of the system in a well-organized manner. The problem of localization of both external and internal edges of the iris of the eye and the elimination of the effects of the upper and lower eyelids is an important issue in the field of research.

The implementation of our algorithm of iris recognition based on simulations of 257 iris images showed us the effectiveness and the performance of our integrated methods. The theoretical analysis of our algorithm of iris recognition, at the level of localization of the external and internal edges of the iris, has shown its practical time efficiency compared to other algorithms.

The introduction of aspects of discrete geometry has efficiency at the level of the processing time of the localization of external and internal edges of the iris with a reduction of 5.5 seconds on average compared to what has been proposed in this field of research. Our method for eliminating the effects of the upper and lower eyelids keeps the essential biometric features of the iris and enables the verification of users with a negligible error rate.

The method of comparison vector between two given gabarits diagonal horizontal vertical approximation (DHVA), issued from our biometric model, gave consistent results more satisfactory than those currently obtained by classical methods. Our classification model of the gabarits DHVA, based on the concept of indexed hierarchical classification by trees and the pretopological aspects, provides a time to search these gabarits better than that obtained by the sequential method.

Our model should be applied in several fields, such as banking, defense, health systems, remote learning, and business operations.

Our perspectives consist of

- Implementing our model in a clinic, a bank, or any other field that requires secure access to sensitive data
- Having an extension of our biometric model based on the iris of the eye, by the integration of a multimodal biometric concept using the fingerprint in order to have the ability to access only in very restricted conditions, such as when the user has ocular problems (e.g., edema)
- Simulating the MAS in the model ICAS for later integration in the real field
- Introducing pretopological aspects at the level of the localization of external and internal edges of the iris
- Integrating the gabarits DHVA in the composition of the private key to decrypt the data

References

Allen, J., Hendler, J., and Tate, A. 1990. *Readings in planning.* San Mateo, CA: Morgan Kaufmann Series in Representation and Reasoning.

Ashtiyani, M., Birgani, P. M., and Hosseini, H. M. 2008. Chaos-based medical image encryption using symmetric cryptography. In *IEEE 3rd International Conference on Information and Communication Technologies: From Theory to Applications (ICTTA 2008),* 1–5.

Azzag, H., Guinot, C., and Venturini, G. 2004. Classification automatique de documents: Application au web. In *11th Rencontre de la Société Francophone de Classification (SFC),* September, France: Bordeaux, 91–94.

Bawany, N. S., Paracha, A. M., and Naz, N. 2004. Role of agents in providing better communication. In *IEEE Student Conference on Engineering, Sciences and Technology,* Pakistan, 169–173.

Belmandt, Z. 1993. *Manuel de prétopologie.* Paris: Hermès.

Bernard, R. 1985. *Méthodologie multicritère d'aide à la décision.* Paris: Economica, 423 pp.

Berta, I. Z., and Vajda, L. 2003. Documents from malicious terminals. *SPIE—International Society for Optical Engineering—Conference Microtechnologies for the New Millennium. Bioengineered and Bioinspired Systems,* 325–326.

Birand, L. 1991. *Systèmes temps réel en ADA: Une approche virtuelle et asynchrone.* Paris: Masson.

Bo, Y., and Qinghua, W. 2000. Agent brigade in dynamic formation of robotic soccer. In *IEEE Proceedings of the 3rd World Congress on Intelligence Control and Automation,* Department of Information and Control Engineering, Shandong University of Building Materials, China, 1: 174–178.

Boella, G., and van der Torre, L. 2005. Role-based rights in artificial social systems. In *IEEE International Conference on Intelligent Agent Technology,* Torino University, Italy, 516–519.

Bonastre, J-F. 2005. Biométrie et multimodalité BIO-MUL. ACI sécurité informatique, laboratoire d'informatique, Université d'Avignon.

Bourdel, L. 1962. *Sang et Tempéraments*. Paris: Fayard.

Bron, A. J., Tripathi, R. C., and Tripathi B. J. 1997. *Wolf's anatomy of the eye and orbit*. London: Chapman & Hall Medical.

Brooks, R., and Connell, J. H. 1986. Asynchronous distributed control system for a mobile robot. In *Proceedings SPIE in Mobile Robots* 727: 77–84.

Cabal, C. 2003. La biométrie. Rapport Assemblée Nationale no. 938 et Sénat no. 335 (Juin).

Cabri, G., Ferrari, L., and Leonardi, L. 2003. A case study in role-based agent interactions. In *IEEE Twelfth International Workshops on Enabling Technologies: Infrastructure for Collaborative Enterprises (WETICE)*, Italy, 42–47.

Cammarata, S., McArthur, D., and Steeb, R. 1983. Strategies of cooperation in distributed problem solving. In *Proceedings of the Eighth International Joint Conference on Artificial Intelligence (IJCAI-83)*, Germany: Karlsruhe, 6 (1): 35–66.

Campbell, S. 2003. Supporting digital signatures in mobile environments. In *IEEE Proceedings of the 12th International Workshops on Enabling Technologies: Infrastructure for Collaborative Enterprises*, 238–242.

Capkun, S., Hubaux, J-P., and Buttuán, L. 2006. Mobility helps peer-to-peer security. In *IEEE Transactions on Mobile Computing* 5 (1): 43–51.

CASIA. 2006. CASIA iris image database version 3. Institute of Automation, Chinese Academy of Sciences.

Chaib-Draa, B., Paquet, E., and Lamontagne, L. 1993. Architecture d'un système intelligent pour les environnements multiagents. In *Actes de l'Équipe de Recherche Intelligence Artificielle de Clermont-Ferrand*, Volcans–IA93, Université Blaise Pascal, Laboratoire Informatique, Université d'Auvergne – IUT d'Informatique, France, 42–55.

Chaib-Draa, B., and Vanderveken, D. 1999. Agent communication language: Towards a semantics based on success, satisfaction and recursion. In *Intelligent Agents V. Agent Theories, Architectures and Language*, ed. Müller, J. P., Singh, M. P., and Rao, A. S., 1555: 363–379. Berlin: Verlag.

Chasse, M. 2002. La biométrie au Québec. Document d'analyse en informatique. Commission d'Accès à l'Information, Québec.

Chassery, J-M., and Montanvert, A. 1991. *Géométrie discrète en analyse d'images. Traité des nouvelles technologies série—Images*. Paris: Hermès.

Chatley, R. 1997. Hybrid deliberative/reactive agents (www.iis.ee.ic.ac.uk/frank/surp99/article2/rbc97).

Chen, J., Zhang, H., and Hu, J. 2008. An efficiency security model of routing protocol in wireless sensor networks. In *IEEE Second Asia International Conference on Modeling & Simulation (AICMS 08)*, 59–64.

Chen, J. J-R., Richard, W-M. L., and Chen, A-P. 2003. A new convertible group signature scheme on the basis of dual complexities. In *Proceedings IEEE of the 37th Annual 2003 International Canadian Conference on Security Technology*, 115–122.

Cho, C-B., Chande, A. V., and Li, Y. 2005. Workload characterization of biometric applications on Pentium 4 microarchitecture. Intelligent design of efficient architecture lab (IDEAL). Department of Electrical and Computer Engineering, University of Florida. IEEE, 11 pp.

Cho, S., Park, J., and Kwon, O. 2003. Gender differences in three-dimensional gait analysis data from 98 healthy Korean adults. *Clinical Biomechanics* 9 (2): 145–152.

Christodoulou, D., and North, S. 2004. Historic maps drawing. Dissertation project, University of Sheffield.

Cite-Sciences. 2005. Biométrie: Est-ce biométrisable? Cité des sciences et de l'industrie, France. www.cite-sciences.fr/francais/ala_cite/expositions/biometrie/wo/biometrisable.html (accessed March 20, 2008).

Conte, R., Miceli, M., and Castelfranchi, C. 1991. Limits and levels of cooperation. In *Proceedings of the Second European Workshop on Modeling Autonomous Agents and Multiagent Worlds (MAAMAW-90)*, Amsterdam, Netherlands: Elsevier Science 2: 147–160.

Crevier, D., and Lepage, R. 1997. Knowledge-based understanding system. *A Survey Proceeding of Vision and Image Understanding* 67 (2): 161–185.

Daouk, C. H., El-Esber, L. A., Kammoun, F. D., et al. 2002. Iris recognition. In *IEEE Proceedings of ISSPIT,* Marakesh, 558–562.

Das, B., and Kocur, D. 1997. Experiments in using agent-based retrieval from distributed and heterogeneous databases. In *IEEE Knowledge and Data Engineering Exchange Workshop (KDEX '97)*, 27–35.

Data Investment Consult. 1999. Electronic commerce: Application & implications. In *Seminar of the Center for Emerging Markets,* Lebanon.

———. 2000. E-banking and the new technology. In *Seminar of the Center for Emerging Markets,* Lebanon.

Daugman, J. 1993. High confidence visual recognition of persons by a test of statistical independence. In *IEEE Transactions on Pattern Analysis* 15: 1148–1161.

———. 2004. How iris recognition works. In *IEEE Transactions on Circuits and Systems for Video Technology* 14 (1): 21–30.

Deluzarche, C. 2006. L'iris. *Linternaute Science.* www.linternaute.com/science/biologie/dossiers/06/0607-biometrie/iris.shtml (accessed February 11, 2007).

Demazeau, Y., and Müller, J-P. 1991. *From reactive to international agents: Decentralized A.I. 2.* New York: Elsevier Science.

Demengel, G., and Pouget, J. P. 1998. *Modèles de Bézier, des B-splines et des NURBS.* Editions ellipses.

Durfee, E. H., and Lesser, V. R. 1989. Negotiating task decomposition and allocation using partial globing planning. *Distributed Artificial Intelligence* 2: 229–243.

Erman, L., Hayes-Roth, F., Lesser, V., and Reddy, D. R. 1980. The HEARSAY-II speech understanding system: Integrating knowledge to resolve uncertainty. *ACM Computing Surveys* 12: 213–253.

Escobar, A. E. 2006. Biometrics. http://www.findbiometrics.com/events.php (accessed December 10, 2006).

Estraillier, P., and Girault, C. 1992. Applying Petri net theory to the modeling analysis and prototyping of distributed systems. In *IEEE Proceedings of the IEEE International Workshop on Emerging Technologies and Factory Automation,* Australia: Cairns.

FaceKey Corporation. 2007. Security by biometrics. The new generation of access control and time and attendance products. San Antonio, TX, 210: 826–8811 (www.facekey.com).

Ferber, J. 1997. *Les systèmes multiagents: Vers une intelligence collective.* Paris: InterÉditions.

Finin, T., Fritzson, R., Mckay, D., et al. 1994. KQML as an agent communication language. In *The Proceedings of the Third International Conference on Information and Knowledge Management (CIKM '94).* ACM Press.

Finin, T., Potluri, A., Thirunavukkarasu, C., et al. 1995. On agent domains, agent names and proxy agents. CIKM Intelligent Information Agents Workshop, Baltimore, MD.

Florin, G., and Natkin, S. 2003. Les techniques de cryptographie. Support de cours, unité de valeur: Systèmes et applications répartis. Paris: CNAM.

Futura-Sciences. 2008. Lettre d'information, © 2001–2008 Futura-Sciences. www.futura-sciences.com (accessed June 13, 2008).

Galliers, J. R. 1988. Theoretical framework for computer models dialog, acknowledging multiagent conflict. PhD thesis, UK: Open University.

Georgeff, M. P. 1983. Communication and interaction in multiagent planning. In *Proceedings of the Third National Conference on Artificial Intelligence (AAAI-83),* Washington, DC.

Gillerm, D. 2007. Les technologies biométriques. http://biometrie.online.fr/techno/empreintes/T-fin_index.htm (accessed January 22, 2007).

Grecas, C. F., Maniatis, S. I., and Venieris, I. S. 2001. Toward the introduction of the asymmetric cryptography in GSM, GPRS, and UMTS networks. In *Proceedings of the Sixth IEEE Symposium on Computers and Communications,* 15–21.

Güneysu, T., Möller, B., and Paar, C. 2007. Dynamic intellectual property protection for reconfigurable devices. In *IEEE International Conference on Field-Programmable Technologies (ICFPT 2007),* 169–176.

Hashem, S. 2000. *L'identification biométrique dans le commerce électronique.* Edition Eyrolles.

Huang, T-P., Weiluo, S., and Chen, E-Y. 2002. An efficient iris recognition system. In *IEEE Proceedings of the First International Conference on Machine Learning and Cybernetics,* 292–395.

Huhns, M. N., and Singh, M. P. 1998. Workflow agents. In *IEEE Internet Computing* 2 (4): 94–96.

Ishikawa, S., Kim, H. S., Inoue, F., et al. 2000. Man–machine collaborative work based on visual communication. In *IEEE Proceedings TENCON 2000,* Malysia: Kuala Lumpur, 2: 321–325.

Jain, A. K., Ross, A., and Prabhakar, S. 2004. An introduction to biometric recognition. In *IEEE Transactions on Circuits and Systems for Video Technology, Special Issue on Image- and Video-Based Biometrics* 14 (1): 4–20.

Jalix. 2001. IFT: Ressource électronique. Support du cours. (http://www.jalix. org/ressources/miscellaneous/security/_IFT6271/intro.pdf).

Kang, M-M., Park, W-W., and Koo, J-R. 2003. Agent for electronic commerce on the semantic web. In *IEEE Proceedings KORUS 2003 on Science and Technology,* Korea, 2: 360–363.

Kearney, P. 1996. Personal electronics, personal agents. In *IEEE Colloquium on Intelligent Agents and Their Applications,* London, 7: 1–6.

Labrou, Y., and Finin, T. 1994. A semantic approach for KQML—A general purpose communication language for software agents. In *Third International Conference on Information and Knowledge Management.* ACM Press.

———. 1997. Comment on the FIPA '97 agent communication language. University of Maryland, Baltimore.

Lanctot, B. 1997. La sécurité informatique. Module du cours *santé et sécurité.* Canada: Ecole Polytechnique Montréal.

Lee, L., and Grimson, W. 2002. Gait analysis for recognition and classification. In *Proccedings of the International Conference on Automatic Face and Gesture Recognition,* 148–155.

Lee, S-E., Shin, S-H., Park, G-D., et al. 2008. Wireless sensor network protocols for secure and energy-efficient data transmission. In *IEEE 7th Computer Information Systems and Industrial Management Applications (CISIM '08),* 315–319.

Lesser, V. R., and Corkill, D. D. 1983. The distributed vehicle monitoring tested: A tool for investigating distributed problem solving networks. *AI Magazine* 4 (3): 15–33.

Lim, S., Lee, K., Byeon, O., et al. 2001. Efficient iris recognition through improvement of feature vector and classifier. *ETRI Journal* 23 (2): 61–70.

Mahmoudi, D. 2000. Biométrie et authentification. Corporate Information and Technology Department, Swisscom, AG (http://ditwww.epfl.ch/ SIC/SA/publications/F1000/fi-sp-00/sp-00-page25.html).

Malone, T. W. 1990. Organizing information processing systems: Parallels between human organizations and computer systems. In *Cognition, computation and cooperation,* ed. Zachary, W. W., and Robertson, S. P., 56–83. New York: Ablex.

Mannino, M. V. 2004. *Database design—Application development and administration.* New York: McGraw–Hill.

Marie-Claude. 2003. La biométrie: Souriez vous êtes filmés. Documentation de Nicolas Six, France (http://souriez.info/article).

Markovie, M. 2007. Data protection techniques, cryptographic protocols and PKI systems in modern computer networks. In *IEEE 14th International Workshop on Systems, Signals and Image Processing, and 6th ERASIP Conference Focused on Speech and Image Processing, Multimedia Communications and Services,* 13–24.

Martial, F. V. 1990. Interactions among autonomous planning agents. In *Proceedings of the First European Workshop on Modeling Autonomous Agents in Multi-Agent Worlds (MAAMAW-89),* ed. Demazeau, Y., and Müller, J-P. Decentralized AI. Amsterdam, Netherlands: Elsevier Science Publishers B.V.

Mayfield, J., Labrou, Y., and Finin, T. 1995. Desiderata for agent communication languages. In *Working Notes of the AAAI Spring Symposium Series,* Stanford University, 122–127.

Meng, H., and Xu, C. 2006. Iris recognition algorithms based on Gabor wavelet transforms. In *Proceedings of the 2006 IEEE International Conference on Mechatronics and Automation,* China: Luoyang. IEEE, 1785–1789.

Meskaoui, M. 2005. *Système multiagent pour un réseau Diffserv: Modèles comportementaux et plate-forme: Intelligence dans les réseaux.* Paris: Hermès Lavoisier, 205–230.

Michael, J. P. 2002. Applications, graphics and security: Exploring visual biometrics. *IEEE* 22 (4) (http://www.wildcrest.com).

Microsoft. 2002. Microsoft official curriculum. Designing a secure Microsoft windows 2000 network. Document manual, Microsoft.

Mintzberg, H. 1979. *The structuring of organizations.* Englewood Cliffs, NJ: Prentice Hall.

Miyazawa, K., Ito, K., Aoti, T., et al. 2005. An efficient iris recognition algorithm using phase-based image matching. In *Proceedings of the IEEE International Conference on Image Processing,* Japan, 2: 49–52.

Molineris, M. 2006. La biométrie: Les enjeux industriels—La sécurité obtenue—Les applications actuelles. Mémoire de l'examen probatoire: Informatique. CNAM Centre Régional Associé D'Aix-En-Provence.

Montain, B. 1999. *Groupe sanguin: Clé de votre caractère.* Éditeur Trédaniel Guy.

Mordini, E. 2005. Point de départ d'un débat Européen sur l'éthique de la biométrie. Le projet ETIB.

Müller, J. 1996. Negotiation principles. In *Foundation of distributed AI,* ed. O'Hare, G. M. P., and Jennings, N. R., 211–230. Chichester, England: John Wiley & Sons.

Nam, D-H., and Park, S-K. 2001. Adaptive multimedia stream service with intelligent proxy. In *IEEE Proceedings of the 15th International Conference on Information Networking (ICOIN' 01),* Division of Information and Computer Engineering, Ajou University, Korea: Suwon. 291 pp.

Narote, S. P., Narote, A. S., and Waghmare, L. M. 2006. An automated iris image localization in eye images used for personal identification. In *IEEE Advanced Computing and Communications* (ADCOM), India, 250–253.

Neo Diagnostico. 2008. Laboratoire d'expertise ADN, spécialiste en analyses génétiques. info.fr@neodiagnostica.org (accessed March 20, 2008).

Noisette, T. 2005. L'usage de la biométrie tend à s'élargir. Actualités ZDNET (July), France.

Odubiyi, J. B., Kocur, D. J., Weinstein, S. M., et al. 1997. SAIRE—A scalable agent-based information retrieval engine. In *Proceedings of the First International Conference on Autonomous Agents,* Marina Del Rey California, 292–299.

Otto, A. 2004. Cryptographie. Dictionnaire informatique (Dico Info) (accessed November 16, 2008).

Padiou, G., and Sayah, A. 1990. *Techniques de synchronisation pour les applications parallèles.* Toulouse: Cepadues—Editions.

Parnes, P., Synnes, K., and Schefström, D. 1999. A framework for management and control of distributed applications using agents and IP-multicast. In *Proceedings of the Eighteenth Annual Joint Conference of the IEEE Computer and Communications Societies (INFOCOM '99)* 3: 1445–1452.

Parunak, H. V. D. 1996. Applications of distributed artificial intelligence in industry. In *Foundation of distributed AI*, ed. O'Hare, G. M. P., and Jennings, N. R. Chichester, England: John Wiley & Sons.

Perronnin, F., and Dugelay, J-L. 2002. Introduction à la biométrie: Authentification des individus par traitement audio-vidéo. *Traitement du Signal* 9 (4): 253–265.

Phillips, P. J., Martin, A., Wilson, C. L., et al. 2000. An introduction to evaluating biometric systems. *IEEE Computer* 33 (2): 56–63.

Portejoie, P. 1991. Planification en univers mono-agents (définitions, concepts, objectifs, exemples, limitations): Introduction à la planification en univers multiagents. In *Actes de l'équipe de recherche intelligence artificielle de Clermont-Ferrand, planification et coopération dans les systèmes à bases de connaissances*, France, 1.

Postnikov, M. 1981. *Leçons de géométrie: Géométrie analytique*. Moscow: Editions Mir.

Potel, M. J. 2002. Applications—graphics and security. In *IEEE Exploring Visual Biometrics* 22 (4) (http://www.wildcrest.com).

Pujolle, G. 2003. *Les réseaux.* Paris: Eyrolles.

Rasmussen, J. 1986. *Information processing and human–machine interaction: An approach to cognitive engineering* (North-Holland series in system science and engineering). Amsterdam: Elsevier Science.

Reuters. 2006. Know. Now. In *Capital Markets Workshop*, Holy Spirit University of Kaslik, Faculty of Business Administration, Lebanon.

Robert, W. I., Bradford, L. B., and Delores, M. E. 2005. Effect of image compression on iris recognition. In *IEEE Instrumentation and Measurement Technology Conference (IMTC 2005)*, Ottawa, Canada, 3: 2054–2058.

Rosistem. 2001. Biometric education. Automatic identification seminar (AIM), Japan. http://www.rositerm.ro (accessed December 16, 2006).

Saporta, G. 1990. *Probabilités—Analyse des données et statistique.* Paris: TECHNIP.

Schonberg, D., and Kirovski, D. 2004. Iris compression for cryptographically secure person identification. In *Proceedings of the Data Compression Conference (DCC'04). IEEE* 459–468.

Shannon, C., and Weaver, W. 1948. *The mathematical theory of communication.* Urbanna: University of Illinois Press.

Smith, R. G., and Davis, R. 1980. Frameworks for cooperation in distributed problem solving. *IEEE Transactions on Systems, Man, and Cybernetics* SMC-11 (1): 61–70.

Stallings, W. 1999. *Cryptography and network security*, 3rd ed. Englewood Cliffs, NJ: Prentice Hall.

Tian, Q., Li, L., and Sun, Z. 2006. A practical iris recognition algorithm. In *Proceedings of the 2006 IEEE International Conference on Robotics and Biometrics*, Kunming, China. 5: 392.

Tian, Q-C., Pan, Q., Cheng, Y-M., et al. 2004. Fast algorithm and application of Hough transform in iris segmentation. In *Proceedings of IEEE Third International Conference on Machine Learning and Cybernetics,* Shanghai, 7: 3977–3980.

Tomko, G. 1996. Chiffrement biométrique. *18th Conférence Internationale de Protection de la Vie Privée et des Données Nominatives du Canada.*

Uludag, U., Pankanti, S., Prabhakar, S., and Jain, A. K. 2004. Biometric crypto-systems. *Proceedings of IEEE* 92 (6).

Vatsa, M., Singh, R., and Gupta, P. 2004. Comparison of iris recognition algorithms. In *ICISIP,* Department of CSE, India. *IEEE* 354–358.

Vongkasem, L., and Chaib-Draa, B. 1999. ACL as a joint project between participants: A preliminary report. In *First workshop on agent communication language (ACL '99),* Sweden: Stockholm, 235–248.

Wildes, R. P. 1997. Iris recognition: An emerging biometric technology. In *Proceedings of IEEE* 85 (9): 1348–1363.

Wilson, L. F., Burroughs, D. J., Kumar, A., et al. 2001. A framework for linking distributed simulations using software agents. In *Proceedings of the IEEE* 89 (2): 186–200.

Winograd, T., and Flores, F. 1986. *Understanding computers and cognition: A new foundation for design.* New York: Ablex Publishing Corp.

WIPO (World Intellectual Property Organization). 2008. Method and device for the virtual simulation of a sequence of video images.

Xu, G., Zhang, Z., and Ma, Y. 2006. Improving the performance of iris recognition system using eyelids and eyelashes detection and iris image enhancement. In *Proceedings of the 5th International Conference on Cognitive Informatics,* China, 2: 871–876.

Xuong, N. H. 1992. *Mathématiques discrètes et informatiques: Logique mathématiques informatique.* Paris: Masson.

Yang, J., Stiefelhagen, R., Meier, U., and Waibel, A. 1998. Visual tracking for multimodal human computer interaction. In *Proceedings of the Conference on Human Factors in Computing Systems (CHI-98): Making the Impossible Possible.* New York: ACM, 140–147.

Yoo, J., Hwang, D., and Nixon, M. S. 2005. Gender classification in human gait with SVM. *Advanced Concepts for Intelligent Vision Systems* 3708: 138–145.

Yu, C., Wu, H., and Wu, M. 2004. A complete model for collaborative virtual environment integrating extended fuzzy-timing petri nets with role and agent technology. In *IEEE International Conference on Systems, Man and Cybernetics,* China, 6: 5503–5508.

Yu, C., Wu, M., and Hou, H. 2004. Role based and agent oriented modeling technology for hybrid avatar and virtual actor in collaborative virtual environment. China: *SMC* 6: 5497–5502.

Zhou, T., and Harn, L. 2008. Risk management of digital certificates in ad hoc and P2P networks. In *IEEE Canadian Conference on Electrical and Computer Engineering (CCECE 2008),* 325–330.

Index